Praise for Bob Berman's

The Sun's Heartbeat

"A deeply enjoyable book.... Berman comes across as the world's most enthusiastic science teacher, writing infectiously about how humans went from the geocentric days of Aristotle to the current heliocentric understanding.... It's hard not to enjoy."

—Mark Berman, *Washington Post*

"*The Sun's Heartbeat* offers a compelling and surprisingly playful history of the Milky Way's most famous star—from the alternately brilliant and misguided theories of the ancient Greeks, to the modern-day discoveries that would make Arthur C. Clarke and Stanley Kubrick blush."

—Jacob Sugarman, *Salon*

"Berman's pitch-perfect book goes a long way to answering the questions you thought were too dumb to ask, but it does much more than simply provide facts. Berman is a master storyteller, whose passion and enthusiasm for astronomy has served the public well for decades.... Read this and you will never look at the Sun in the same way again."

—Michael Brooks, *New Scientist*

"This might be the last book you ever read—afterward, you can't help but stare, in wonder, directly into that fiery ball in the sky. From ancient Sun worship to the latest in Sol science, Bob Berman makes *The Sun's Heartbeat*."

—Sam Kean, author of *The Disappearing Spoon*

"A quick, smart, and colorful biography of 'yon flaming orb.'"

—*Kirkus Reviews*

"Berman directs your attention to our neighborhood ball of nuclear fire, telling its story with charm and wit.... He makes a compelling case for putting on a wide-brimmed hat, stepping outside, and giving a second thought to the star that illuminates and powers our planet."
—Shannon Palus, *Discover*

"An informative look at the life and death of the star on which our lives depend.... The author's enthusiasm for science shines through."
—John Gribbin, *Wall Street Journal*

"The Sun—Bob Berman always capitalizes the name of his beloved—is 'the most awesome thing around,' he exults. After reading this entertaining popular-science page-turner, it's tough to disagree.... Many of the concepts in this book are mind-bending, but Berman repeatedly employs humor to bring the star-struck reader back to earth."
—Duncan McMonagle, *Winnipeg Free Press*

"Bob Berman's lively new book... offers a number of fresh perspectives—on the everyday miracle of sight, on the importance of spending time in the Sun, on the reality of global climate change.... Berman comes to the rescue of parents (and grandparents) of hyper-inquisitive youngsters everywhere."
—Roberta Conlan, *AARP*

"From Coronal Mass Ejections to eclipses to sunspots to sun burns, Bob Berman writes of all things solar with inescapably contagious delight. *The Sun's Heartbeat* reminds us that the star we take for granted, the star that we accept with little thought, is both central to our lives and wonderfully complex. Berman's book, fun and informative, shows us how to see our flaming orb in a new light."
—Bill Streever, author of *Cold*

"Bob Berman's *The Sun's Heartbeat* glitters and skips with the joy and excitement of science at its best. He explains things I always wondered about without diminishing the star-gazer's sense of awe."

—Mark Kurlansky, author of *Salt* and *Cod*

"Berman extols the Sun's aesthetic effects—most spectacular of them, the total solar eclipse, rivaled, perhaps, by the northern lights, with rainbows as second bananas. An engaging consciousness-raiser that entertains as it informs about our neighborhood nuclear furnace."

—Gilbert Taylor, *Booklist*

"Berman shakes readers out of a complacent understanding of his subject with startling facts conveyed in companionably witty prose.... Making this common sight mysterious again, he reminds us of questions we had forgotten to ask."

—Margaret Quamme, *Columbus Dispatch*

"Bob Berman's look at our sky's most important star is as dazzling as the Sun itself. The man can *write!* He is one of those rare authors whose prose is as delightful to read as it is pleasing to learn from. Having read this book, I will never be able to regard the Sun—nor the future beneath it—in quite the same way."

—James M. Tabor, author of *Blind Descent: The Quest to Discover the Deepest Place on Earth*

"We won't take the Sun for granted any longer if astronomy popularizer Berman...has anything to say about it....'Everything about the Sun is either amazing or useful,' Berman writes, and then proves it, without a doubt."

—*Publishers Weekly*

Also by Bob Berman

Biocentrism
(with Robert Lanza, MD)
Shooting for the Moon
Strange Universe
Cosmic Adventure
Secrets of the Night Sky

The Sun's Heartbeat

And Other Stories from the Life of the Star That Powers Our Planet

BOB BERMAN

BACK BAY BOOKS
Little, Brown and Company
NEW YORK BOSTON LONDON

Back Bay Books / Little, Brown and Company
Hachette Book Group
237 Park Avenue, New York, NY 10017
www.littlebrown.com

Originally published in hardcover by Little, Brown and Company, July 2011
First Back Bay paperback edition, July 2012

Back Bay Books is an imprint of Little, Brown and Company. The Back Bay Books name and logo are trademarks of Hachette Book Group, Inc.

Illustrations by Alan McKnight

Library of Congress Cataloging-in-Publication Data
Berman, Bob, author.
The sun's heartbeat : and other stories from the life of the star that powers our planet / Bob Berman. — First edition.
p. cm.
Includes bibliographical references and index.
ISBN 978-0-316-09101-5 (hc) / 978-0-316-09099-5 (pb)
1. Sun — Popular works. I. Title.
QB521.4.B47 2011
523.7—dc22 2010044207

10 9 8 7 6 5 4 3 2 1

RRD-C

Printed in the United States of America

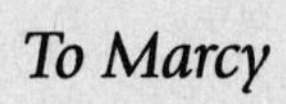

To Marcy

Contents

The Sun's Heartbeat

Introduction

Every fifteen years or so, a bright incoming comet intrigues the media, and the juicy topic "danger from space" gets a brief resurrection. But peril from the sky is never very far from our collective mind-set. The paranoid among us, not content to worry about cholesterol, wonder whether we will soon get clobbered by a giant meteor. It's not completely far-fetched, since a dinosaur-ending impact did indeed change Earth and a dramatic comet explosion over Siberia in 1908 did radically reduce that region's already low property values. Nonetheless, only one astronomical body truly rules Earth and always has. It's the nearest star—the one bearing the second-shortest name of any celestial object.

Most people probably think they know the Sun well enough. It's bright; it's hot; it can bestow a handsome tan or a painful burn. What else do we need to know? Reacting with our deep skin tissue, sunlight produces the vitamin D that is the greatest anti-cancer agent ever discovered. There. Done. Why not leave all other Sun knowledge to geeks and scientists?

But we can scarcely ignore the only outer space object that controls absolutely everything in our neck of the cosmic woods. Especially when it keeps changing unexpectedly, in ways that affect us personally.

The Sun today is measurably different from the way it was a year ago and especially ten years ago. The weather outside your window, right now, is the result of the Sun's altered behavior. Indeed, it's been acting so strangely that a special NASA conference was convened in 2008 to try to figure out what is going on. Meanwhile, researchers continue to follow a fascinating, powerful, 11-year activity pattern that has intrigued scientists for 250 years.

The Sun rules your life and mine, and how much we pay for things, and where we may ultimately choose to live, and whether our next flight will divert from its intended polar route, and even whether humans will ever colonize other worlds. Its mutating behavior manipulates us much more than our own volcanoes and earthquakes. The Sun influences crop yields, global temperatures, ocean currents, and human moods. Its most recent cycles have even altered the realities of climate change.

The Sun's transition to dominance in my own thinking took three decades, a languid metamorphosis I attribute to youthful slow-wittedness. I occasionally met with solar researchers during my seventeen years as a monthly *Discover* columnist; still highlight the Sun every year as the astronomy editor of *The Old Farmer's Almanac;* and I happily keep abreast of new discoveries as part of my work as contributing editor and columnist for *Astronomy* magazine. But it was the dark, starry heavens with their magical cobalt galaxies that pulled at my spirit, and the Sun was the night's greatest enemy. Even so, it kept tugging at my sleeve: stars are impossibly distant dots, but this one is *right here,* offering its secrets, as deep as they can be, in exquisite detail.

Probably the turning point for me was a series of Sun-related assignments. First I was eclipse leader at half a dozen mind-bending totalities, then an Alaska aurora borealis lecturer for three winters. Without even consciously intending to, I was altering my college courses — making the Sun take up more and more of the curriculum for my students. Its hypnotic attraction had snuck up the way it had millennia ago for the ancient Mayans, Greeks, and Egyptians.

Then, since 2000, what had previously been a trickle of new Sun discoveries became a flood. And a flotilla of six amazing solar-dedicated spacecraft were almost routinely providing revelations. How many of your friends know that there's a "sun inside the Sun"? Or that a bizarre, newly found zone beneath the solar surface, the *tachocline,* is solely responsible for its violence? Or that we just experienced the oddest solar cycle in more than two hundred years—which has apparently influenced global warming in a major way? Or the scary stuff, like an all-star government panel that urgently warned that a particular kind of solar storm could effectively destroy our power grid at any time, with repairs to the tune of one to two *trillion* dollars?

Is it really mostly the Sun's changing brightness—and not human activity—that's altering our temperature? Is my own *Old Farmer's Almanac* correct in predicting, ostensibly using scientific reasoning, that a global *cooling* is in the cards?

How can we create a valid public policy regarding climate change if the culprit is not ourselves but the nearest star? Or are such arguments mere rationalizations and cop-outs, a fog created by a deliberate or ignorant misreading of today's cutting-edge solar science? And what about the medical advice to cover our skin when we venture out—has this drained our blood of its vital cancer-fighting vitamin D? And what should we make of the Sun's eternal hand-to-hand combat with our world's magnetic field; its complete power over our solar system and satellites; the billions of particles it powers through our brains?

It's time to find out. But before we reach such destinations, it is just as compelling to see how that nearest star works as it creates all that is interesting and animated in our neighborhood. It's a stellar voyage with surprises at every turn. And most of all, it's a story of laughable errors, egotistical battles, brilliant inspirations, genius technology, and nature at its quirkiest. This is the story of our long ascent to understanding that terrifying, life-giving ball of nuclear fire.

CHAPTER 1

Yon Flaming Orb

That lucky old Sun has nothing to do
But roll around heaven all day.

—Haven Gillespie, "That Lucky Old Sun," 1949

EVERY DAY, a ball of fire crosses the sky.

This bizarre reality was so flabbergasting in ancient times that people could relate to it only with worship. The Aztecs and Egyptians were far from alone in regarding the Sun as a god. The Persians, Incas, and Tamils (of southern India) also elevated the Sun to the center of their spiritual lives. It was giver of life, holder of all power, an enigma beyond all earthly mysteries. You either bowed to it or just shut up and gawked.

Today it's just "the Sun." Familiarity is the enemy of awe, and for the most part people walk the busy streets with no upward glance. In fact, one of the common bits of advice about the Sun is that we shouldn't look at it.

At its most elemental, the Sun is the sole source of our life and energy. We know this. And when the Sun occasionally pops into the mainstream news, like an old dog briefly roused from a nap, we give it a few moments' attention. Recently, medical headlines announced that we've been hiding from the Sun excessively in an understandable desire to avoid skin cancer. The long-standing advice is now changing: it's better to get too much sunlight than too little. That's because sunlight-mediated vitamin D prevents far more cancers than are produced by going too far the other way and staying pale. Sunlight is not just good for us; it can save our lives.

This is news worth sharing. But, in truth, everything about the Sun is either amazing or useful. Beyond the peculiar history of its discoveries by a motley collection of geniuses, it directly or indirectly affects our lives, our health, our emotions, and even our dollars.

BECAUSE WE MUST START SOMEWHERE, we might spend a few minutes with the simplest solar truth, the one noticed even by hamsters and trilobites: the Sun is bright. Meaning, it emits lots of energy. We know it gives off ultraviolet. This is what burns us at the beach, and should we get too much of it, we might contract a fatal melanoma. Every 1 percent increase in a person's lifetime UV exposure brings a 1 percent boost in her chances of being victimized by that deadliest of skin cancers. Indeed, Australia's high melanoma rate has its simple origin in that region's characteristically sunny skies.

We also need no expert to tell us that heat, those long infrared energy waves, gushes from the Sun. We feel it on our faces even when we're behind glass and thus receiving no ultraviolet at all.

But what does the Sun emit most strongly? Heat? Ultraviolet? Gamma rays? What?

Few would guess the answer. It is green light. That is the Sun's peak emission.

The green dominance shows up in rainbows and in the spectrum cast on a wall by a prism, where green always seems to be the brightest color. Yet the Sun itself does not appear green. This is because our eyes were designed by nature so that when the Sun's primary colors—green, red, and blue—strike the retina together, the mixture of light is always perceived as white. White means we're getting it all.

When looking at grass, we might reasonably conclude that the botanical world—and the chlorophyll that fuels it—loves sunlight's abundant emerald component. But leaves, plants, and grass actually do not like green at all. They feed mostly on the Sun's blue light and also absorb its red wavelengths. Grass reflects away the green part of sunlight, which is why we see that color in lawns. Counterintuitively, the only thing we ever perceive is an object's *unwanted* solar wavelength.

Our eyes perceive objects solely because particles of light or photons are bouncing off them. We've been designed to see the energy the Sun emits most strongly. It would be pointless to be sensitive to gamma rays, for example, because the Sun sends us scarcely any; they're not bouncing off the objects around us. Whether we're searching for a cheeseburger or underwear, we'd see nothing at all if our eyes were scanning for gamma rays. Instead, we see the most abundant solar wavelengths.

This gives us a Sun bias. When we look at the night sky, we're comfortable with planets, which stand in the light of the Sun. And we feel good under a starry sky, since all stars give off the same type of light, composed of precisely the same colors, that the Sun does. We essentially scan the universe through the Sun's eyes. Our retinas and brains were built to "Sun specs." We're not quite grapefruit, which seem to mimic the Sun's very appearance, but the Sun is sort of a mother-father figure as far as our construction goes.

Since the Sun's energy peak is green light, that's also the color we perceive most easily. In deepening twilight, when light turns murky and colors fade, we still see grass as green, even though red sweatshirts and violet flowers have all turned to gray—the first step toward the color blindness we experience at night.

When we're camping far from artificial lights under a full moon, the world looks green-blue. It's such a familiar experience, we don't stop and say, "What's going on here? Why should a white moon make everything turquoise?"

Photographers, artists, and cinematographers are very aware of this strange effect and happily exploit it whenever they wish to portray an outdoor night scene. Since the moon's surface is really just a dull mirror made of finely powdered sand, moonlight is sunlight whose brightness is cranked down 450,000 times. So cinematographers often shoot a sunlit scene, then use a filter to block out the warm colors and also cut down the light a few f-stops to make it dimmer. Voilà: the illusion of moonlight. But why exactly is the moonlit world cyan?

In daylight, our retinas perceive all colors with a peak sensitivity to yellow-green. But in low light, our perception shifts blueward. Now our greatest sensitivity is to the color green-blue, while hues at the edges of the spectrum, such as red and violet, vanish and become gray. This perception change at low light levels is called the *Purkinje shift,* named for the Polish scientist who first noticed it. (Jan Purkinje was also the first to suggest that fingerprints be used in criminal investigations.) "Purkinje shift" sounds so cool that I try to say it as much as possible, even when it's not appropriate. Thanks to the varying sensitivities of the retina's three types of cone-shaped color receptors, when light is fairly dim, we perceive nature's greens but not its oranges and reds. This explains why most municipalities now paint fire trucks green. Gone is the traditional red, which cannot be seen well at night. It's also why green was chosen for all US interstate highway signs—all forty-six thousand miles of them.

The oddities of the Sun's light and color are limitless. Why, for instance, does the Sun produce violet but no purple, and why do we humans not see the real color of the daytime sky? But looming over all these Sun-related questions is the big one, the one asked by children since earliest recorded history: what makes the Sun shine?

This was surely one of the top ten most baffling questions for twenty-five thousand generations. Even ancient scriptures claiming wisdom or omniscience, such as the Vedas and the Bible, didn't dare touch it, because nobody had a clue. We have known the answer for just one human lifetime. The underlying physics is now well understood and has yielded unexpected side oddities, such as the strange particles called *neutrinos,* which are created in the Sun but then magically change form as they fly through space. By-products of the Sun's turning its own body into light, twenty trillion neutrinos whiz through each of our heads every second.

That the Sun is such a tireless power source is not surprising if we recall that nature uses material objects to store huge amounts of energy. When this energy is converted, the process unleashes an almost inconceivable fountain of light and heat. This "mass changes to energy" idea went unsuspected even by great thinkers such as Plato and Isaac Newton. Until a century ago, all hot, self-luminous objects were simply regarded as forms of fire, the act of burning. Everyone assumed that the Sun was burning, too.

Nineteenth-century scientists already knew that Earth completes one orbit per year, and when they figured out our planet's distance from the Sun, the simplest physics let them calculate the mass of the star holding us in its grasp. It turns out the Sun weighs the same as 333,000 Earths. If its body consisted entirely of coal, the resulting fire could keep going for just two thousand years, and never mind where it would get the oxygen needed to sustain combustion. That means that if the Sun first ignited when the pyramids were built, the fire would have been out by the time Christ was born. Obviously, something was wrong with this picture. No matter the fuel, the Sun just couldn't be burning. This enigma, staring us in the face daily, plagued the greatest minds.

It took Einstein's "energy and matter are the same thing" idea a century ago to reveal the true process. Grasping the significance of $E = mc^2$ a few years later, in 1920, the British astronomer Arthur Eddington correctly proposed that vast energy is released when hydrogen's proton — the heavy particle in its nucleus — meets and sticks to a proton from another hydrogen atom.

Eddington was criticized because the required temperature for proton fusing was not thought to be available in stars such as the Sun. He replied, "I am aware that many critics consider the stars are not hot enough. They lay themselves open to an obvious retort; we tell them to go and find a hotter place."

Touché, nice retort, but actually he was right. The fusing together of hydrogen nuclei is what makes the Sun shine.

By exchanging the burning-coal idea for the notion of nuclear fusion, science was really trading an amazing wrong idea for an amazing right one. Given the total power emitted by the Sun, which delivers nearly a kilowatt of energy to each square yard of Earth's sunlit surface every second, and the formula $E = mc^2$, it's easy to calculate how much of the Sun's body gets continuously consumed and turned into light. The truth is a little disconcerting: the Sun loses four million tons of itself each second.

This is not some mere abstraction. If we had a giant scale, we'd find that the Sun actually weighs four million tons less every second. So its energy isn't "free." It will pay a price for such a profligate output in the form of a series of dramatic events that will change the Sun into something unrecognizable from the star it is today. The main milestones, which will happen 1.1 billion years, 6 billion years, and 9 billion years from now, can be observed with fascination elsewhere in our galaxy when we observe other suns living out the middle and end of their lives.

"LET THERE BE LIGHT." For centuries, this was easy to say but impossible to understand. Light, color, the Sun's genesis, and the puzzles of its energy production have been pondered since the earliest humans acquired brains large enough to be tormented. But even these mysterious fundamentals are perhaps one-upped by a single intriguing solar property utterly unsuspected by the ancients: the Sun has a heartbeat. A pulse.

As first observed in the seventeenth century, the Sun alters its appearance and energy output in eleven-year cycles and reverses its magnetic polarity every twenty-two years. Although the Sun's heartbeat is longer than the human variety, it's just as dependable and as crucial. Depending on what part of the Sun's period we're in, weather gets cooler or warmer, Earth's atmosphere thickens up, and the yield of wheat crops appears to respond, along with the average price of bread around the world. Ocean currents shift markedly.

Such effects on Earth make them worth our focus, especially since recent solar cycles have displayed extremely strange behavior—such as 801 days without a single sunspot—that has not been seen since the 1600s and that has resulted in powerful earthly influences. It even made some researchers wonder whether Earth was on the verge of returning to something like the frightening seventy-year period during early colonial times when the Sun's heartbeat simply stopped.

One of the consequences then was a significant cooling of our planet. You might guess the natural next question: might the Sun be stepping in now to unwittingly help counteract global warming? Can this explain why the runaway carbon dioxide in our atmosphere has so far produced only a wimpy 1.5°F rise in Earth's surface temperature over the past century?

But recent solar behavior has been both stranger and stronger even than this. The Sun's modern somnambulance has unfolded while our planet's own magnetic field has been steadily weakening. This double whammy translates into an inability to stoutly block incoming cosmic rays from the depths of the galaxy. The Sun's rhythm since the year 2000, and especially its super-low ultraviolet output from 2006 to 2009, unseen until now by any living researcher, has created the darkest night skies since the Roaring Twenties. It has boosted our air's carbon-14 content. It's flattened the global warming curve. It's chilled our upper atmosphere, which is 74°F colder than in any normal sunspot minimum, making it so thin and compressed that it has affected the lives and deaths of satellites. And the Sun's own magnetic field is only half as strong as it was a mere twenty years ago. The bottom-line question is, what exactly is going on, and how is it affecting us?

We should really start at the beginning, rewinding back through human history—which has a kind of heartbeat of its own—to explore the star that, while doing so much else, hovers at the center of our lives.

CHAPTER 2

Genesis

The sun itself sees not till heaven clears.

—William Shakespeare, Sonnet 148

THE SUN'S BIRTH was accompanied only by silence. There were no onlookers. The nearest planets were light-years away, and that was a good thing. Like the modern art its looping, gassy tendrils resembled, the apparition would have been best appreciated from a great distance.

The process by which a boring cloud of plain-vanilla hydrogen gas becomes a blinding ball of white fire is epic in purpose and scale. The result, a stable star such as the Sun with a fourteen-billion-year life span, destined to create puppies and pomegranates, certainly deserves its own holiday. Yet no nation celebrates the Sun's birth. We do, theoretically, honor its existence each Sunday.

In practice, most use that time to sleep as late as possible and thus minimize any awareness of it.

What if we had been present at the moment of the Sun's creation, 4.6 billion years ago? Anthropomorphically enough, the Sun came from the union of two parental sources. It had a true mother and father, each boasting a distinct personality.

The maternal side was a vast womb of thin hydrogen gas created 379,000 years after the big bang. This nebula was probably ten million times bigger than the Sun it produced, which is like a football stadium birthing a single apple seed on the fifty-yard line.

Into this cold, placid, Siberian near emptiness came roiling violence approaching from a distance. A massive blue star had blown up in our galaxy's Perseus spiral arm. The supernova's detritus, the paternal side of the solar DNA, included new substances utterly unknown in the nostalgic days of the big bang, nine billion years earlier. The meeting and mixture of these two disparate materials produced the star that was destined to be ours.

Although stars are always born in the same fashion, they end up very different from one another. Some become colossal orange spheres that would dwarf our planet the way a spherical tank twenty-five stories high would compare to the period at the end of this sentence. Such globes expand and contract like an anesthesiologist's oxygen bag—surprisingly lively motion for something so enormous. Others reach their AARP years as tiny white balls crushed solid, harder than a diamond and smaller than Los Angeles.

Within a large cloud of hydrogen and helium—a nebula—any lump or extra bit of density exerts its own gravity and pulls in surrounding gas. During this contraction, an initial "barely there" random motion evolves into a leisurely spin, and as the blob shrinks, like a ballerina pulling in her arms, the rotation speed increases.

Gravitational collapse always produces heat, so the center gets ever hotter. Meanwhile, the nebula shapes itself into a ball. Stars and planets are spheres because these have the smallest surfaces of any geometric shape. No part is farther from the center than any other.

When you were little and played with clay, you could patty-cake it into a thin sheet, like pizza dough. If you wanted to color the large surface when it hardened, you needed lots of paint, especially considering both sides. Or, instead, you could roll the clay between your palms into a small ball that could be painted with a single brushful. You learned then that balls have small surface areas. Nature learned that, too, early on. Any object with enough material — it takes only one-hundredth the mass of our moon — has enough self-gravity to pull itself inward until it reaches the smallest 3-D shape, a sphere. Across the cosmos, only very low-mass objects such as asteroids avoid having round bodies.

So we have a spinning gas ball with a hot center. Now a critical event — nuclear fusion — either happens or fails to happen.

With enough gas and thus enough self-gravity present, the spinning ball's core compresses to be white-hot. Heat simply means "moving atoms," so each hydrogen atom is now hurled at its neighbor with such force that their protons stick together: *fusion.* It's a form of alchemy, except that instead of gold, the fusing of hydrogen's protons creates useless helium, good only for party balloons or making our voices sound like a Munchkin's.

But proton fusing also releases heat and light. This is why, when you ask a physicist what makes stars shine, he says either "fusion" or, if he thinks you can handle it, "the proton-proton chain."

THE MOMENT FUSION BEGINS, a star is officially born. The process releases such enormous energy that it's self-sustaining. And this energy is immense. In each second, the Sun produces and emits the same energy as six trillion Hiroshima atomic bombs. This fusion will continue on its own for as long as hydrogen fuel remains — and in the case of the Sun, which weighs two octillion (2 followed by 27 zeros) tons, the fuel supply is no problem.

If the original hydrogen cloud is too skimpy, its self-gravity produces a ball with insufficient heat to get fusion started. The result is a brown dwarf. It never "shines." It never creates its own light. But it's still too hot to touch. In a way, the planet Jupiter is like that. Its core is the hottest place in the solar neighborhood, outside of the Sun itself, but it would have needed eighty times more mass than it has to ignite and become a star.

But what if it had ignited? Then ours would be a double-star system. Such binaries are so common that they make up half the stars in the universe. They run the gamut from two stars nearly touching each other, pulled tidally into football shapes and whirling crazily around their common center of gravity every few hours, to twins so far apart that millions of years must elapse before they complete a single mutual orbit.

The weight of the collapsing nebula determines everything. Too little mass and fusion never happens. Just enough and the low core pressure makes fusion proceed slowly: the star shines dimly as a cool red dwarf. With only one-thousandth the solar mass, such lightweight stars emit a feeble orange glow. These are the street dogs of the galaxy; they're everywhere. Forty-three of the fifty stars nearest to Earth are of this dim, ruddy variety, which astronomers, without any apparent logic, call *type M*. These "Methuselah suns" use fuel so sparingly that they outlive everything else.

At the other extreme are massive stars. When these implode and ignite, gravity is so powerful that the core fusion is virtually a runaway, proceeding crazily, almost like a bomb: superheat creates a blue dazzle. Such blinding lighthouses are rare but appear disproportionately in the night sky because we see them from thousands of light-years away. But their fuel is spent quickly, with no thought to the future. These stars don't last long.

It was just such a profligate, live-for-the-moment blue star that fathered our own Sun.

Astronomers classify these distinct star families with simple capital letters. There are only seven major star types—as if to retain the ancient classic obsession with that number, as in "seven seas" or "Seven Wonders of the World." In many cultures, both six and eight were deemed unlucky and thus were avoided, which is why you rarely hear tales of the "six dwarfs" or the "Eight Sisters."

Originally, the seven star categories were labeled A, B, and so on, according to how much hydrogen emission was in their light. But the sequence soon got reshuffled according to the stars' temperature. For more than a century now, astronomers everywhere have known these star types—O, B, A, F, G, K, M—as intimately as chefs know their fundamental seasonings or geologists the basic kinds of rock. The letters are a continuum from the bluest, hottest, most massive stars (types O and B) to the reddest, coolest ones (type M). In between are creamy white, moderate stars of type G (like the Sun) and yellow or orange type K. There are no green stars in the universe.

These wildly disparate stellar life stories play out in real time in the twinkling night. Rigel, every stargazer's familiar friend in Orion's foot, is just like our Sun's father. It is a type B star, 17 times heavier than the Sun, which makes it shine 65,000 times brighter. It lives happily for only 7 million years by changing hydrogen to helium, the most normal and efficient fusion behavior. As Rigel's hydrogen gets depleted, it will increasingly "burn" helium, which creates carbon and oxygen, a process that will last only 700,000 years.

And things keep going downhill. The next step, changing some of its newborn oxygen to silicon, will buy Rigel only one solitary year. The final metamorphosis of silicon to iron will keep the star alive for a single additional day. Iron is the end of the line. Making atoms heavier than iron *costs* energy rather than producing it.

Then, out of fuel and down on its luck, with no outward-pushing power streaming from its core, the star will finally give in to gravity. The sheer weight of all its layers will make the star collapse on itself, rapidly upping the temperature until the whole thing blows as a supernova — the most intense brilliance nature ever creates.

The core will remain as a tiny, crushed ball twelve miles wide, spinning crazily. The rest of the star will be hurled outward two thousand times faster than a rifle bullet. Two million planet Earths' worth of newly minted star stuff — all that fresh-baked oxygen, silicon, iron, and the rest — will fly across space in a flash brighter than a million Suns.

But more than that, a supernova's incredible heat fuses new ultraheavy atoms. Indeed, the only possible way for nature to create iodine, lead, uranium, and every other element heavier than iron is in a supernova's cauldron. So what we now have flying through space is absolutely all of the ninety-two natural elements, a big change from the mere hydrogen and helium and a dash of calming lithium the star started with. The massive star has been a factory, creating every element, including the oxygen we are breathing this very moment.

A supernova's brilliance fades in a year or two, but the material keeps going. It would be a shame for it to go to waste.

Encountering a primordial nebula, the supernova detritus pushes that gas into dense filaments, creating knots of collapsing gas balls and a litter of new stars. These so-called second-generation stars are partially made up of heavier elements from the now deceased blue star. They are what astronomers call *metal-rich.*

Some of these second-generation stars are high-mass, blue O and B suns that go through their own life cycles in sped-up double time. When *these* blow up into supernovae, an even richer explosion spreads across the galaxy. When this enhanced material contacts yet another virgin gas cloud (these nebulae are almost everywhere), it forms third-generation stars, which are the latest models to date. Such stars, and any leftover matter that condenses into planets around them, are rich in oxygen and carbon. They become playgrounds for nature's creative experimentation.

Still with me? If so, you've earned the punch line: our Sun is a third-generation star.

The planets orbiting it reflect this. That we have iodine in our thyroid glands proves that our bodies were fashioned from supernova material. The iron in our blood came from the cores of two previous star generations. The Sun gives off a bit of peculiar yellow light from fluorescing sodium vapor, an element inherited from its father, the type O or B blue star. It got copious hydrogen, which will supply it with life-sustaining fuel for billions of years to come, from its mother, the pristine nebula gas. Their marriage vows were exchanged 4.6 billion years ago. Since then, our galaxy has spun around twenty times, and we cannot pinpoint what, if anything, is left of the nebula that was our nursery. Nor do we see any trace of the supernova that was the other component of our genesis.

The residual, gaseous afterbirth of the Sun's (and our own planet's) origins has been diluted and lost over time. But the mechanism of the birth stares us in the face. Seen through even a one-dollar spectroscope, sunlight contains the fingerprints of its complex composition. Its high metal content is conclusive evidence that, unlike so many other stars in the universe, the Sun was not forged from primordial material alone. The universe harbors many ongoing mysteries, from the big bang to the nature of consciousness, but how the Sun was born is not among them.

Our own parents were the first generation of humans for whom the Sun's origin no longer required the desperate guesses of philosophers or the invoked mysteries of theologians. The glow of gaseous metals contributes to the sunlight reflecting off the white feathers of gulls flying overhead and the faces of children playing in the yard. The Sun's genesis illuminates the clouds above and the rippling waves of the sea below. It is all around us. In fact, in many ways, it *is* us.

CHAPTER 3

A Strange History of Seeing Spots

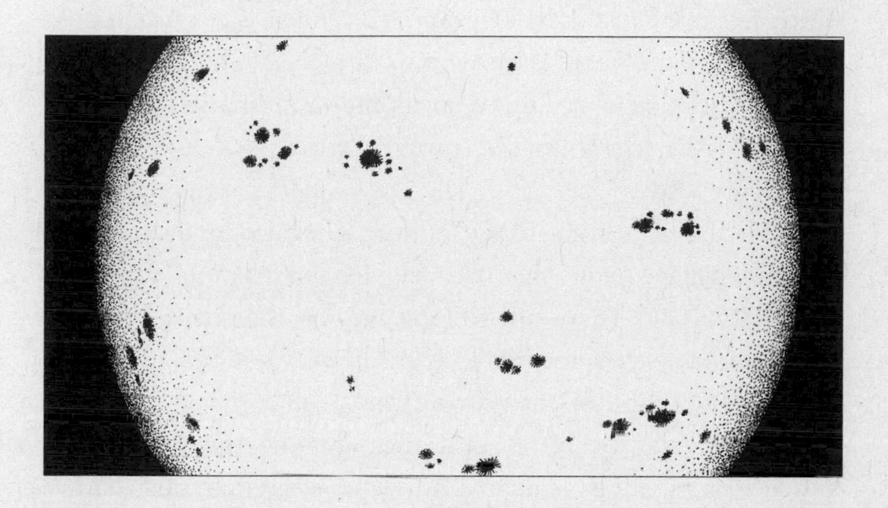

I hold this opinion about the universe, that the Sun remains fixed in the centre, without changing its place; and the Earth, turning upon itself, moves round the Sun.

—Galileo, Letter to Cristina di Lorena,
Grand Duchess of Tuscany, 1615

I, Galileo, son of the late Vincenzo Galilei, Florentine, aged seventy years, arraigned personally before this tribunal, and kneeling before you, . . . swear that I . . . [will] abandon the false opinion that the Sun is the center and immovable, and that the Earth . . . moves.

—Galileo, at his inquisition, June 22, 1633

AS THE FIRST humans acquired tools and an appreciation of minimalist cave art, they turned their attention to improving their lives and understanding the cosmos. *Homo erectus* erected the first blazing fire 500,000 years ago. After burgers went from raw to medium-well, a truly long time elapsed before the next human milestone: the bun. The first planting of grains and other crops, which occurred just 12,000 years ago, ended our million-year low-carb diet and freed us from being hunters. No longer plagued by the frustration of trying to sneak up on animals with bigger ears and faster legs, humans started staying put. Our nomadic days were ending.

After the beginning of agriculture, the next milestone was written language. This was cuneiform, invented by the Sumerians around 3400 BC. That's not so long ago. We thus have less than six thousand years of records to let us know what bygone people thought about the Sun or anything else.

The Egyptian Museum in Cairo—among the world's top must-see destinations—is a wonderland of hieroglyphs that, at first glance, look like a kindergartner's idea of animal portraiture. When they were finally deciphered in the mid-nineteenth century ("Aha, I see! It's snake before stork except after fish!"), the inscriptions revealed how central the Sun was to daily life. Here was a god no one treated lightly.

The early Egyptians, of course, did not limit themselves to fine motor skills. When it was finished, the Great Pyramid at Giza, forty stories tall and weighing 6.5 million tons, was the most precisely aligned structure in the world, with each side exactly aimed toward the four cardinal directions with an accuracy of more than one-tenth of a degree. Such massive monuments were thus more than mausoleums; they doubled as aids in solar and seasonal reckoning.

Numerous early civilizations noticed that the Sun's rising and setting positions shift regularly, and so they placed markers to keep track of the changes. You can do this, too. Starting the first day of winter, the solstice (December 21), watch where the Sun sets from your least obstructed window. As the days and weeks pass, the Sun will keep setting farther to the right. On some particular evening, it might go down behind some "monument," such as a neighbor's chimney or a distant telephone pole. On June 21, the summer solstice, it will set as far to the right as possible. Then for the next half year, it will move to the left, retracing its steps, again setting at all those horizon points it hit during the first six months. Thus, each spot on the horizon marks a sunset on two occasions each year. The solstices alone, at the left and right terminals, host the Sun only once.

In Manhattan, the setting Sun stands like an orange ball at the end of every numbered street on May 31 and again on July 11—"Manhattanhenge"—though rush-hour crowds seem disinclined to gather and stare. In Salt Lake City, whose roadways are far more perfectly aligned with the cardinal directions, the Sun dramatically and blindingly sets at the end of every street on March 20 and September 23—the equinoxes. If you were an obsessive-compulsive member of an ancient sky-watching culture and had a favorite community field, you'd certainly feel compelled to erect a set of stones to mark at least the extreme sunset positions, the two solstices.

The Mayans, the Egyptians, and those who built Stonehenge are usually said to have used the Sun to tell time. But "time" is abstract. It has no independent existence outside of being a human tool of perception. You cannot pick it up and analyze it, like cottage cheese. Indeed, Thomas O'Brian, the head of the National Institute of Standards and Technology's Time and Frequency Division—which builds and runs our sophisticated atomic clocks—told me over lunch that he has "no idea what time is." Neither did the ancients. Early cultures cared not about "time," but about synchronicity and practical need: they knew that certain sunrise and sunset positions told them to plant crops or prepare for a likely stretch of scorching heat and low rainfall, which had previously happened when the Sun set at around the same place on the horizon.

The Egyptians represented the rising Sun as the god Horus, always depicted as a falcon. His job was to drive away darkness, and he never failed in that era when the aches and injuries from performing repetitive tasks went unrecognized. But even he played second fiddle to Ra, the chief Sun god, who also had the head of a falcon, but with a disk above it. It's possible that the "wings" protruding from this disk may have been an esoteric depiction of the coronal streamers seen during a total eclipse.

Like those who came to power later, such as the Inca rulers and Japanese emperors who used the same ploy, the pharaohs claimed to be direct descendants of the Sun, thus assigning to themselves a pedigree it would be patently unwise to challenge. Imagine today, when we routinely belittle our leaders, if the president of the United States convinced us that he was descended from the Sun. Could he not get bills passed more easily—especially those involving solar energy?

The rulers at the height of the ancient Greek and Roman Empires lacked this degree of chutzpah. But Sun gods remained. To the Greeks, it was Helios, who each day rode across the sky on a chariot pulled by four horses we can only assume were clad in a fireproof material like asbestos.

Sun gods of various cultures often spilled over into other cultures, sometimes with name changes. For instance, Baal, worshipped by the Phoenicians, was also venerated by many Israelites. This was sufficient competition for the One God of the Bible that Baal put-downs appear periodically in its pages. "The priests of Baal were in great numbers," disdainfully recounts 1 Kings 18:19.

"Will you steal, murder, commit adultery, swear falsely, [and] make offerings to Baal?" asks Jeremiah 7:9. In practice, although the Sun god Baal was also the chief object of worship among the Canaanites, the name Baal started to be used for all manner of deities, which drove the Old Testament–writing rabbis bonkers.

Equally bonkers were the Sun myths recounted in numerous texts without editorial comment, no matter how ludicrous they might seem to us now. A famous Native American legend explains that a spider carries the Sun from a cave into his web each day. Surely every rebellious Iroquois teenager thought, "Yeah, Grandpa, right. A spider brings the Sun into his web. Sure."

Chinese mythology says that long ago, the Sun gave birth to ten suns, which all lived in a mulberry tree until they flew into the sky. This, of course, made Earth way too hot, so an archer named Yi shot nine of them down. Apropos of nothing, the remaining Sun has a crow permanently living inside, which sometimes nibbles away pieces of it. Even to this day, the traditional Chinese symbol for the Sun is a red disk with a crow inside it.

Did anyone actually believe these stories? Some tales, no matter how dubious, were repeated for generations, but wiser heads naturally asked higher-level questions, such as whether Earth goes around the Sun or vice versa, and whether the Sun is a flat disk of fire, like burning kerosene on a plate, or, instead, a fiery sphere with no earthly analog whatsoever.

Great thinkers from less than a dozen countries came up with good answers and changed people from stargazers who gawked dumbly at the sky to those who still gawk, but on an advanced level. We cannot be too cocky these days, since our present grasp of the structure and origins of the cosmos remains so clearly incomplete. But the process of learning about sky objects started five thousand years ago with the Babylonians, Sumerians, Egyptians, Chinese, and later the Mayans, who accurately chronicled solar, lunar, and planetary cycles. Oddly enough, many other notable contemporary cultures—the Hebrews, Celts, Romans, and Japanese come to mind—brought us no astronomical advancement whatsoever.

Through it all, in the six centuries before Christ, the Greeks alone went far beyond merely observing and chronicling celestial patterns and rhythms. They came up with correct *explanations.* It's a pity these individuals' names are not universally known. They truly advanced our knowledge about the Sun and its relation to Earth, and even more or less figured out these two bodies' correct distances, sizes, and motions. Yet outside of a few chance school assignments, they are now as forgotten as the Broadway headliners of the Gay Nineties.

Thales of Miletus (ca. 624–546 BC) was one founder of modern physical science. He thought that everything was made of water (true of living things, anyway) and that Earth was a disk floating in a huge ocean. Okay, but he was also the first to accurately plot the path of the Sun across the sky. He predicted the eclipse of May 28, 585 BC, which supposedly halted the battle between the Lydians and the Medes. Because of this eclipse, historians have been able to accurately pinpoint the date of this event, the earliest dated battle in history.

Pythagoras (ca. 580–500 BC) created the famous theorem we reluctantly studied in high school, but he didn't do as well as an astronomer. He believed Earth is a sphere, but he thought it sat, unmoving, at the center of the universe, with the Sun going around it. At least he got the "Earth is a sphere" part right, even if a large majority continued to think otherwise for another two thousand years.

Anaxagoras (ca. 500–428 BC) correctly believed that objects on Earth and in the heavens are made of the same substances. He rightly said that the moon reflects light from the Sun, rather than glowing on its own.

Heracleides Ponticus (ca. 388–315 BC) was light-years ahead of his time. He was the first person known to propose that since Mercury and Venus stay so close to the Sun, they might orbit it, and that Earth might rotate on an axis. These ideas are absolutely correct, even if ignored at the time. How ironic that this most perspicacious of the early Greeks is also among the least known today.

Aristotle (384–322 BC) is still arguably the most famous of the ancient Greeks, but he held back science for the next two thousand years with his geocentric model of the universe, which went unquestioned until the time of Galileo and which bewilderingly became church doctrine. Later on, questioning Aristotle meant getting burned at the stake. A few of his other writings were correct. For example, he believed Earth is spherical, not flat.

Hipparchus of Nicaea (ca. 190–120 BC) discovered the twenty-six-century wobble of Earth's axis called *precession,* and he created the first accurate star catalog, dividing stars into six magnitudes of brightness, a system that is still used today. He also determined the length of a year to within six minutes, even though he imagined he was timing the Sun's annual orbit around Earth instead of the other way around.

Despite all these accomplishments, two other Greeks, Aristarchus of Samos (ca. 310–230 BC) and Eratosthenes of Cyrene (ca. 276–194 BC), probably deserve to sit atop Mount Olympus as the greatest of the great. Their deductions were not only dead-on correct but also eighteen centuries ahead of everyone else's.

ARISTARCHUS WAS A mathematician and astronomer—the first person known to write and preach that the Sun is the center of the solar system and that Earth orbits around it once a year. Of course, it never pays to be significantly ahead of your time. Even for the freedom-loving Greeks, Aristarchus went too far. Cleanthes, the leader of the Stoics, tried to have him indicted on the charge of impiety for claiming that Earth doesn't sit as the stately, immobile king of the cosmos, but rather does a lively circus two-step by orbiting the Sun while also spinning on an axis. To his contemporaries, that must have seemed as comical as suggesting that Earth pats its head while rubbing its belly. But when we justly praise Galileo and Copernicus for advocating the heliocentric theory, let's remember who said it first, seventy-two generations earlier.

Aristarchus even made a stab at figuring the relative distance from Earth to the Sun and moon. Mathematically and logically, the idea is simple. When the moon precisely reaches its half-moon phase, the time when it should theoretically be located at right angles from the Sun, it does not actually stand 90 degrees from the Sun in the sky, but less. This geometry can easily be used for calculating where the Sun must be located, just as a friend at a known distance, whose face is exactly half-illuminated, tells you where the room's lightbulb must be found.

Without the benefit of a telescope, however, figuring where exactly the moon sits the moment it is half-lit has a typical error of a few degrees — after all, the moon moves its own width each hour and seems just as perfectly "half" at 7:00 PM as at 8:00 PM. It's not an easy judgment. Indeed, Aristarchus thought that when the moon was half, it stood 87 degrees from the Sun in the sky. Using that figure, he determined that the Sun must be eighteen times farther from Earth than the moon. And since they both appear to be the same size in the sky, it must mean that the Sun is physically eighteen times larger. This would make it several times bigger than Earth, and this general truth — that the Sun is at least bigger than Earth — is perhaps what convinced him that the Sun lies at the center of motion. A smaller body should logically orbit a larger one, not the other way around.

We know today that the half-moon actually hovers 89¾

degrees from the Sun, which places the Sun 400 times farther away than the moon, and therefore makes it 400 times bigger, which amounts to 109 times larger than Earth. Aristarchus had no way to precisely measure the correct angle, but he absolutely had the right idea. And, more than for anything else, he wins the cigar for his heliocentric model.

Aristarchus lived to be eighty, enjoying to his last days the slow pace of his native island of Samos. When he was thirty-five, a fellow genius, destined for an identically long life, was born in the city of Cyrene, now in Libya. Eratosthenes was brilliant, especially in his specialties of math and mapmaking, and it was he who coined the word "geography." Eratosthenes's great mind was recognized by all who knew him, and few were surprised when, at age forty, he was appointed by Ptolemy III to be in charge of the great library at Alexandria. Today that world famous repository of learning has been rebuilt, and after it reopened in 2002, I was thrilled to go there, though disheartened to discover that a foreigner must pay sixty dollars to get a library card. Still, *a library card from the library of Alexandria!* Almost irresistible. One wonders what it cost back in the library's classical heyday, and for that matter, what they charged for an overdue papyrus.

Eratosthenes made a noble but futile attempt to calculate the distance to the Sun, too, but he remains most famous for being the first to accurately determine the size of Earth. And he did it without ever setting foot outside Egypt.

He was able to accomplish this seemingly impossible task because he recalled that on the day of the summer solstice, the Sun stands perfectly straight up for those who live in the city the Greeks called Syene, now Aswan. (Confusing things further, that large town on the Nile was known as Swenet by the Egyptians.) The straight-up Sun was not lost on Syene's residents, who on that day could see sunlight shining to the very bottom of their wells. A precisely overhead Sun was the kind of cool thing that nowadays would become a tourist draw, complete with postcards. Eratosthenes was also very aware that the Sun is never even close to straight overhead as seen from his library in Alexandria. On the solstice, it misses the zenith by more than 7 degrees, or 14 Sun widths, or about one-fiftieth of a full circle.

Today any high school junior could set up a simple proportion to compare the 7-degree difference in the Sun's angle between Syene and Alexandria with the 360 degrees in a full circle (meaning a ratio of 1:50), to obtain the unknown size of Earth. Here were three "knowns" and one "unknown." Since Eratosthenes was both a mapmaker and a mathematician, the job was a piece of cake for him, too, even if no one had ever previously thought of it. The only tricky part was knowing the distance between Alexandria and Syene. Once he knew that, he'd just have to multiply it by fifty, and bingo: the circumference of our world.

Eratosthenes used the speed of camels and riverboats along the Nile and the time it took people to make the trip between Syene and Alexandria. You wouldn't think that would be accurate enough to work, especially since any error was going to be multiplied fifty times over, but it did. He concluded that the cities were 5,000 stadia apart. The stade was equal in length to the average stadium of the time, very nearly one-tenth of a mile. Multiplying by fifty, he deduced that Earth was 252,000 stadia around, which happens to be the correct value within 1 percent.

Unlike other geniuses who were belittled or burned at the stake, Eratosthenes was quoted and supported by notable writers for centuries. (So much for the silly but oft-repeated notion that Columbus was ahead of his time in believing that Earth is round—though he was indeed ahead of his unschooled peers.) The only thing later in doubt was the exact length of the stade Eratosthenes used. Two thousand years in the future, scholarly second-guessers pointed out that the common Attic stadium was about 185 meters, and that would make Earth's circumference come out to 29,000 miles, with a diameter of 9,200 miles, which is 16 percent too big. But if we assume that Eratosthenes used the 157.5-meter Egyptian stadium (and why not—he was, after all, the librarian in Alexandria), his 252,000 stadia means that Earth is 24,466 miles around and 7,788 miles wide. This figure comes wonderfully close to the known diameter of 7,926 miles today. For that, Eratosthenes definitely wins the giant stuffed bunny.

TWENTY-FOUR YEARS AFTER THE BALDING, lifelong bachelor Eratosthenes died around his eightieth birthday, another soon-to-be-famous Egyptian was born in that same city of Alexandria. Claudius Ptolemaeus, soon and forever after known as Ptolemy, thought deep thoughts but ended up being wrong about nearly everything. In his *Almagest,* Ptolemy turned his back on the brilliant works of Eratosthenes and Aristarchus and instead supported the idea of a geocentric universe, which became gospel for the next seventeen hundred years.

Ptolemy embraced Aristotle's simplistic idea that everything is composed of the elements earth, water, air, and fire and that each element has its natural place and motion. Fire's natural place is "up," so it makes sense that flames and smoke always rise. Water and earth prefer being "down," so items such as hamsters and pottery, composed of those elements, fall easily. It was hard to argue with.

Moreover, he said, the rules change beyond Earth. From the moon and beyond, all things are faultless and immutable, and their natural motion is neither up nor down but circular, since the circle is God's perfect shape (because it's beginningless and endless). To him, this explained why the Sun and moon move around us in circles (actually, they don't), and it nicely jibed with the religious notion that God is "up there" in the perfect realm.

Tidy. But, of course, as wrong as serving Chicken McNuggets with white wine. So some Greeks nailed the truth, while others offered circular reasoning. Unfortunately, these latter views were the ones passed on and revered for the next two thousand years, thus keeping this garbage in schools and churches until the Renaissance, when Copernicus, Kepler, and Galileo finally prevailed. Even then, getting it right was a Sisyphean struggle, since anyone who contradicted the church's Ptolemaic doctrine of perfection above was asking for serious trouble down here below.

In the late sixteenth and early seventeenth centuries, great peril lurked for anyone who preached that the Sun is mutable. The Italian scientist Giordano Bruno not only embraced the Sun-centered model of Aristarchus (and the more contemporary Copernicus), but he went so far as to suggest that space goes on and on, and that the Sun and planets are just one of many such systems in the cosmos. What would he say next? feared the church. That humans were not the only intelligent beings God created in the universe? For his prescience, Bruno was burned at the stake in 1600.

Spots on the Sun, which we now know to be the most obvious sign of solar changes that affect our lives and our world, were also the first bearers of information about the Sun itself. Of course, they came and went, and they seemed like blemishes, and thus were incompatible with the established notions of celestial perfection and changelessness. Then as now, however, a lot depended on your connections. With enough higher-ups in your Rolodex, you might just get away with some low-level blaspheming.

Such was the situation during the first few years after the invention of the telescope in 1608 by a Dutch spectacle maker. In 1610 and 1611, no fewer than five observers could make seemingly legitimate arguments that they were the very first to telescopically discover sunspots. Then as now, controversy gets noticed, and the world quickly became eager to confer great retroactive celebrity on whoever proved to be the first.

Johannes Fabricius, who lived in either Holland or Germany, depending on the map or time in history you choose, found sunspots on March 9, 1611, and excitedly watched them with his father for days on end through their telescope, until their eyes were damaged. Alas, good judgment comes only from experience, and experience comes only from bad judgment. These earliest solar observers had no one to warn them that, especially through a telescope, the Sun can quickly cook retinal cells.

Fabricius hurriedly published his discovery that very autumn — inarguably the first European book to discuss sunspots. Or maybe not inarguably. The Italian painter Raffael Gualterotti had already published his own detailed description of his naked-eye observation of a sunspot in 1604, which certainly trumped all the telescope users' claims of priority.

Meanwhile, English telescope maker Thomas Harriot, fresh from a voyage of discovery cruise to Roanoke, Virginia, with Sir Walter Raleigh, started recording sunspots in December 1610. Since Harriot mailed his descriptions and drawings to colleagues, thereby documenting them, the British to this day regard him as the first discoverer of sunspots. Harriot had actually been the navigator and go-to science person on two voyages to the New World. Spending more than a year there, he was one of the very few who bothered to learn the Native Americans' language and much about their customs. Back in England, he wrote a bestseller titled *Briefe and True Report of the New Found Land of Virginia*, which made him a celebrity and earned him a wealthy patron. That's how he could afford to buy a telescope even before Galileo had one. One bit of irony is that in his book, Harriot praised the natives' bodies by saying they were "notably preserved in health, and not know many grievious diseases, wherewithal we in England are often times afflicted." He attributed this partly to their extensive smoking of tobacco, which he lauded because "it purgeth superfluous fleame and grosse humors and openeth all the pores and passages of the body." Harriot himself took to smoking this miraculous substance and ultimately died of cancer that originated in his nasal membranes.

The true fireworks, however, erupted between Galileo, who likely observed sunspots in 1610 with a twenty-power telescope but failed to document them (even though he later insisted that he had told many friends), and the German Jesuit mathematics professor Christoph Scheiner, who first observed sunspots with a student on a date the student later insisted, to Galileo's pleasure, was March 6, 1611. Actual priority of sunspot discovery has continued to be hotly debated by scholars for the past four hundred years, but at the time this issue created an unbelievably nasty international fight between Galileo and Scheiner that lasted until Galileo died decades later, in 1642.

Scheiner's thorough and intense observations through the spring of 1611 resulted in letters later that year to an influential magistrate, who talked him into publishing his findings under the pseudonym Apelles. Because he was a Jesuit, hiding his true identity was important. After all, he was reporting that the Sun had blemishes, although for years afterward he believed the spots were objects that, like little planets, merely surrounded the Sun and thus did not corrupt it. Three years later, after his book got rave reviews and the public clamored to know who the author really was, a fellow Jesuit proudly announced that it was Scheiner who had first discovered the spots.

Galileo read Scheiner's first sunspot book in January 1612, and for the next two years, the conflict between them raged. It ultimately engulfed the educated classes of both countries and reached such a furor that the genuinely earlier observations of Harriot, who had made 450 meticulous drawings of the spots, and of Gualterotti and Fabricius were essentially ignored.

Galileo and Scheiner published digs at each other that resembled the dialogue of two neurotic Woody Allen characters. Galileo first thought the spots were clouds or a vapor, while Scheiner wrongly clung to the notion that they were not at all associated with the Sun's body. He pooh-poohed Galileo in print by asking sarcastically, "Who would ever place clouds around the Sun?" Galileo publicly replied, "Anyone who sees the spots and wants to say something probable about their nature."

Galileo correctly charted the spots' daily progress across the Sun and also rightly realized that some spots vanished off one side and then reappeared on the other two weeks later. He knew they were on, rather than near, the Sun because they assumed a foreshortened appearance when approaching the Sun's edge.

But Galileo went too far in his criticisms. He wrongly ridiculed Scheiner's later claims that the Sun was not uniformly bright. He also erred in dismissing the German's discovery that the spots moved faster near the solar equator—which made Scheiner the first to realize that, unlike the moon and Earth, the gaseous Sun has a differential rotation: its equator spins completely in twenty-five days, but its polar regions require more than an extra week. Scheiner was also the first to see that the Sun's axis is tilted 7 degrees, causing the spots usually to follow curved paths across its body.

Galileo adored this particular Scheiner discovery. It could serve him well in his crusade to demonstrate one of the Sun-centered theory's subplots—that celestial bodies spin with cockeyed tilts. But there was no way that Galileo was going to credit Scheiner with anything that big, so he merely started claiming that he had discovered it first. "Plagiarism!" shouted Scheiner to all who would listen. And yet Scheiner also tried to maintain the expected public composure of a Jesuit. We can imagine him seething, forcing a sickly smile over clenched teeth. Yes, those were fun times.

For the remainder of his life, Galileo vociferously insisted that he alone had found the first sunspots, and he seldom wasted an opportunity to criticize Scheiner. The Jesuit, bitterly offended by Galileo's attacks, struck back, possibly even using his connections to land Galileo in hot water with the church, which ultimately led to his permanent house arrest and forced recantation—also through clenched teeth.

Anyone whose mistakes have ever been embarrassingly pointed out would have to sympathize with how the brilliant if opinionated Galileo felt when Scheiner prominently published a list of twenty-four of Galileo's errors. The Italian went through the roof, writing, "This pig, this malicious ass, he catalogues my mistakes which are but the result of a single slip-up."

Despite the fire and smoke, both men mightily advanced our understanding of the Sun. The sunspot drawings Galileo published in 1613 were far superior to any others, and he was the first to believe that the spots were part of the Sun itself. But Scheiner correctly observed the Sun's rotation to average twenty-seven days over its entire surface—curiously, the same period as the moon's spin. He was also the first to note that no spots appear near the poles, nor are they precisely on the Sun's equator, but rather in a zone 30 degrees north or south of it—roughly analogous to Earth's subtropical belts. And Scheiner alone discovered that the Sun's dark areas are intimately accompanied by bright ones. He named these brilliant patches *faculae,* Latin for "little torches." These bright spots help explain why periods of maximum sunspots are, paradoxically, when the Sun emits the most energy.

In the end, however, none of these European observers was

even remotely the first to discover sunspots. The ulcer-inducing war between Galileo and Scheiner would be like an argument between two modern bike-riding Kansas teenagers who, coming upon the Mississippi, both claimed to be the first human to discover the river. Three thousand years earlier, on a tablet in the world's largest city, Babylon, an unknown writer described a black spot observed on the Sun on New Year's Day. The Chinese also repeatedly noticed spots on the Sun long before the time of Christ, with the first mention of them coming as early as 800 BC. They kept detailed records of these spots starting in 28 BC. At least one of the ancient Greeks, Theophrastus, as well as the Roman poet Virgil, also kept records of them.

Since the prevailing Western religious belief was of a perfect and changeless Sun, it's no surprise that few Europeans noted sunspots prior to the Renaissance. The laudatory term "risk taking," so popular in childhood education today, carried a different connotation back then that strongly implied it was not at all a good thing. Even so, some cast caution to the wind and wrote what they saw anyway. On December 8, 1128, John of Worcester made a drawing on parchment showing two unequally large, black spots on the solar disk, along with this notation: "There appeared from the morning right up to the evening two black spheres against the Sun." This is a particularly credible report because Korean records give a vivid account of an unusual, bright red aurora five days later. Indeed, the amazing and almost mythical northern lights are among the most dramatic visual echoes of the dynamic connection between our world and the changing Sun.

It's clear that people saw spots on their sacred Sun throughout history. But neither Galileo, Scheiner, nor any other of the newly hatched cadre of dedicated solar observers suspected that just as their own lives were drawing to a close, these same spots would deliver to Earth a lifetime of suffering.

CHAPTER 4

The Heartbeat Stops — and Other Peculiar Events

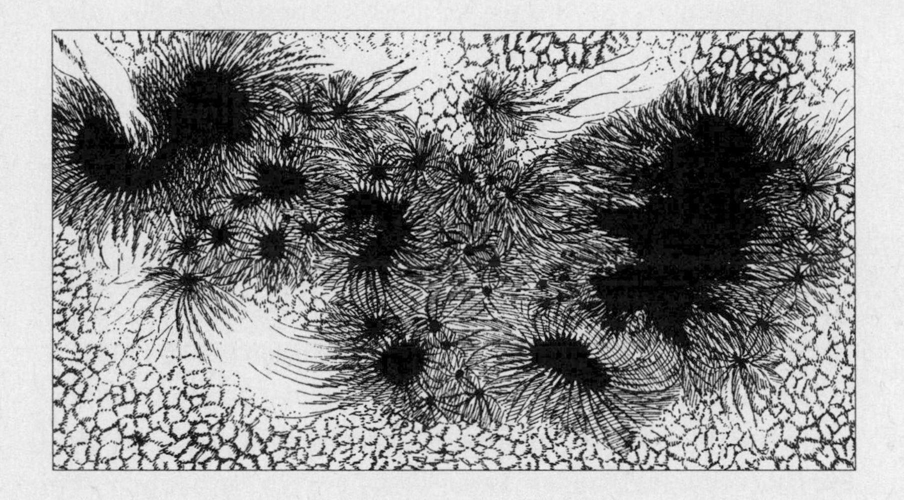

It is also evident, that the whiteness of the Sun's Light is compounded of all the Colours.

—Isaac Newton, *Opticks*, 1704

GALILEO, INCREASINGLY FRAIL and under house arrest, lost interest in sunspots in the 1620s and died in 1642. But his nemesis Christoph Scheiner carried on like a man obsessed. Scheiner's magnum opus, his 784-page *Rosa Ursina*, was finally published in 1630. Despite boasting a hundred drawings and much original information, the book was an unfortunate study in tedium. It was a slow, boring read, and one critic at the time noted that everything worthwhile in it could have been contained in a fifty-page pamphlet. This was the wrong time to publish a Sun book suited to insomniacs. Already the spots were losing their popular interest for several reasons.

First, the number of sunspots was greatly diminishing, as the Sun was now on the cusp of entering an astounding seventy-year period during which its heartbeat — its eleven- and twenty-two-year cycles — all but vanished. Second, no pattern had emerged to hold scientists' interest. Indeed, it was another two centuries before anyone noticed anything but randomness in solar markings, let alone grasped their implications. In science, randomness is normally the kiss of death, like a radio astronomer looking for ET signals and detecting only decades of unrelenting static. The spots displayed no discernible pattern other than moving left to right and avoiding the Sun's poles. They were starting to seem no more scientifically important than the comings and goings of ducks on a pond.

Third, the information stream seemed stuck on "pause." Nothing new was forthcoming. The word on the street was that no one could possibly figure out the Sun's secrets by watching these random ephemeral spots. Who, then, would invest years of observations and create detailed drawings without any hope of a scientific payoff? By the 1620s, Galileo became fond of saying that everything there was to be found about sunspots had already been discovered — and always with the tacit, or not-so-tacit, "by me."

Johannes Kepler, still influential, also could find nothing exhilarating to say about the spots, even when the priority-of-discovery furor was making headlines in 1613. "Like clouds they appear, divide, dissipate, [and] disappear," he wrote. His tone displayed all the excitement of a man waiting for a bus. Kepler was, however, an early believer that the spots were *on* the Sun, not merely hovering or orbiting near it. He described them as being "like slag on the surface of molten metal."

We are fortunate indeed that Scheiner and a few other obsessive-compulsive types kept on diligently observing the Sun despite decades of public and professional apathy. Their more or less unbroken seventeenth-century records have proved to be invaluable, since, unknown to anyone at the time, this was when the Sun was getting truly odd.

THE WEATHER IN A strangely cold Europe got worse after 1645. Crops were failing. Diseases multiplied. The northern lights vanished and became mythical — a bizarre, not-believable apparition that engendered skepticism when old-timers carried on about them, spinning them like exaggerated fishing tales. The aurora borealis's nearly total fifty-year absence lasted until the second decade of the eighteenth century. But nobody connected all these aberrations with the Sun or its lack of spots.

By the mid-seventeenth century, new players had entered the Sun-watching arena. Foremost among them were Johannes Hevelius, who diligently observed from his home-built observatory in what is now Gdańsk; Pierre Gassendi, who carefully watched the Sun from Paris; and Gassendi's best friend, Nicolas Fabri de Peiresc, who had the advantage of living in and observing from the clear skies of the south of France.

But the person destiny placed in the best position to continuously monitor the Sun was John Flamsteed, newly appointed as Britain's first astronomer royal (the position was initially called royal observator). Flamsteed erected and equipped a dedicated solar observatory at Greenwich, plus assembled a team of special observers, mostly scientifically minded clergymen. (Sun watchers at this point routinely used colored-glass filters to safely cut down the solar intensity, probably after their HMOs threatened to withdraw their ophthalmology coverage.) The group monitored the

Sun every clear day, which in England essentially meant never. These clergymen cum scientists, a normal hybrid vocation of the period, sent Flamsteed regular reports by post.

Except there wasn't much to report. In 1676, when Flamsteed noted a sunspot in his records, he wrote that it was "the first I ever saw." He saw one again in 1679, then none until 1684 — an astonishing five-year hiatus. A skimpy few appeared for two or three years, but then the spots totally vanished again between 1687 and 1694. Flamsteed wrote to one of his trusted observers, William Derham, perhaps to assure him that he mustn't feel bad at reporting nothing for so long, "All the stories you have heard of them are a silly romance... to abuse the credulous." It was an assignment akin to asking a Key West lighthouse keeper to keep track of icebergs. The observers must have wondered whether they had been given mere busywork to satisfy their need to feel useful.

In May 1684, Flamsteed wrote to another friend, "Tis near 7½ yeares since I saw one before they have been of late so scarce, however frequent in the days of Galileo and Scheiner." No one knew whether this blankness was normal for the Sun, or whether the spotted Sun was more usual. By 1715, when the spots finally returned — and then soon appeared in great numbers — there had elapsed since the dawn of the Galileo-Scheiner era thirty-five years with spots and seventy years without them, plus a decade of marginal records in which they were apparently infrequent. Who could then say what was customary for the Sun and what was not? Nowadays, of course, things are different. When the Sun went months without a single spot in 2008 and 2009, it threw solar specialists into an uproar. This modern blank Sun was a sight unseen by any living person, smashing a ninety-six-year record for length of time with no spots.

A low sunspot count is one thing. For the Sun to stand totally bereft of them, not just for a few months but for a year or more, is something we now know to indicate a deep underlying disturbance in its magnetic field. Despite continuing observations by Flamsteed and his team—plus Hevelius, Gassendi, Peiresc, and others, most of whom frequently corresponded with Flamsteed's team—no spots whatsoever were seen in 1636 and 1637, nor from January 1645 through December 1651, from 1653 to 1660, from 1676 to 1679, from 1680 to 1688, from March 1689 to May 1695, and from June 1695 until a single one was sighted in October 1700, and then no more until December 1702. Whenever a spot appeared during the second half of the seventeenth century, solar astronomer John (Jack) Eddy noted three hundred years later, it was enough "to occasion the writing of a paper on it." Some even doubted that there ever *were* copious sunspots. After all, they had to balance their own multiyear observations against the chronicles of a handful of early observers, some of whom had been going blind from improper filtering, and all of whom were now dead and thus unavailable for comment. These late-seventeenth-century astronomers probably alternated between skepticism and envy when reading Scheiner's account from fifty years earlier: "From the year 1611 to 1629 I *never* found the Sun quite clear of spots, except for a few days in December, 1624; at other times I was able to count 20, 30, and even 50 spots upon the Sun at a time." Most grudgingly accepted that the solar surface was, for unknown reasons, very different now.

If Flamsteed and his contemporaries had had any inkling of how truly bizarre the Sun's state was, they might well have thought to link it with the strange natural conditions unfolding around them. These were strange indeed. Europe was unusually cold—compared with its future weather during the twentieth century, as well as the first twelve hundred years after the birth of Christ, when Greenland was actually green.

Starting around 1620 and accelerating after 1645, the cold was brutal and almost unrelenting. Viking colonies in the North Atlantic were abandoned. Northern Italy and Crete experienced "famines, droughts, and multiple bad winters." In London, the Thames increasingly had unprecedented periods of ice, and finally experienced multiple winters when it froze solid for months on end, a phenomenon that has not been seen since. Many Scottish and Italian lakes that normally remained liquid in winter became solid as well. Cold pouring down from the poles froze the ocean around Iceland, destroying the cod fishing and decimating the human population with famine.

By 1675, Robert Hooke, secretary and astronomer of the Royal Society, described London as "wild cold." The twentieth-century English climatologist H. H. Lamb said that temperatures in the Faeroe Islands averaged nine degrees below their hundred-year mean. In Sweden, failed crops were becoming routine, with winters and frosts enduring much longer than usual. The cold and frequent food shortages sufficiently weakened the population of Europe that diseases became epidemic. It was an awful time.

By the 1680s, Venice's canals were freezing solid in winter, something no one alive had ever witnessed. Amsterdam's canals were freezing, too. The Norwegian fishing industry failed, as some species could neither survive the plunging sea temperatures nor apparently migrate elsewhere.

Chinese records indicate similar hardships. Two independent chroniclers of weather in China covering six centuries described the period 1640 to 1720 as "cold." And when the Kangxi emperor completed his sixty-one-year rule in 1722, he declared, simply, "The climate has changed."

Japan seems to have fared better during this period, but the British colonies in North America, like the Scandinavian colonies in Iceland and Greenland, knew only hardship. Early cold autumns and brutal winters were frequently accompanied by drought. The winter of 1667–1668 was called "the severest of the seventeenth century." In Boston in 1692, a Reverend Pike described conditions as "the terriblest winter for continuance of frost and snow, and extremity of cold, that ever was known; 31 snowstorms from November 20 to April 9; with the Charlestown [Massachusetts] ferry frozen for six weeks; and a 42 inch snow depth reported at Cambridge." No doubt those recently arrived from Europe were cursing travel agents for their unrealistic balmy portraits of the New World.

Nearly unremitting warfare also became rampant on the European continent, as if the miserable natural conditions fomented discontent. Men of sufficient means threw up their hands and fled across the Channel to establish new lives in England as refugees — deserters, actually. Apparently, quality of life was slightly less miserable there. Of course, it is speculative to link social ills with natural ones, and yet it is not totally far-fetched to see how the Sun's odd behavior could, in a series of despondent, all-too-human steps, become echoed by our own. If the Sun goes on a tremendous bummer, how can we not follow its lead?

The widespread hardship, as well as the absence of sunspots, lasted one human lifetime, from 1645 to 1715. Oddly enough, this cessation of the Sun's heartbeat coincides almost perfectly with the rule of the French "Sun King" Louis XIV (r. 1643–1715). When the Sun King died of gangrene at the age of seventy-seven, the spots immediately came back, and the weather dramatically improved. Apparently, the world could have either a fully functioning Sun or a Sun King, but not both.

Spots or no, it would be an understatement to characterize the Sun as a mystery to the science luminaries of the seventeenth and eighteenth centuries. Into this void came speculation masquerading as fact, a torrent that flooded the literature and was often taken seriously when expressed by the most respected minds of the time. We, of course, have the advantage of hindsight. Who can say which of today's widely repeated seeming truisms about the big bang, string theory, or the universe's origins will seem ludicrous a century hence? Today's farthest-out mysteries are commonly seen as abstruse and possibly insoluble. Desperation makes the public more forgiving, and ideas are rarely ridiculed when *any* hypothesis seems better than none at all. Perhaps the same parameters applied back then when the subject was the Sun.

BY THE MID- TO LATE SEVENTEENTH CENTURY, the Sun's size and rotation period were more or less known, and everyone already realized that its existence was vital for life. The post-Renaissance public, increasingly literate, wanted its geniuses to fill in this skeletal outline and provide some real information about that nearest star. Frustrated teachers fervently wished to do more than shrug when questioned by students. As the telescope's invention passed its fiftieth and then its one hundredth anniversaries, with celestial discoveries that filled volumes, it seemed bizarre that the only facts known about the Sun were the few basics of spots, tilt, and rotation expressed back in 1630 in Scheiner's *Rosa Ursina*.

This vacuum was filled by two centuries of speculation, some of it patently ludicrous, even if written by scientists revered in every other way. The 1856 standard textbook, *Olmsted's School Astronomy,* informed students that, according to no less an authority than William Herschel, discoverer of the planet Uranus, the Sun was inhabited by humanlike creatures who lived on its surface. How could they survive the heat? Simple: High above the Sun's meadows were two layers of clouds. The outermost layer radiated all the fierce fire. The innermost layer blocked this inferno like a hazmat sheet of fiberglass and thereby protected the citizenry below. According to Herschel, observers could even see holes in the clouds and glimpse the cool, dark surface below. Those holes were the sunspots!

Herschel's double-cloud scenario had many adherents. The bright faculae, they thought, were regions where the luminous layer was thicker and hence more brilliant. The gradations in the darkness of each sunspot—from the innermost, inkiest zone, called the *umbra,* to its surrounding aureole, or *penumbra*—were various thicknesses in the luminous layer surrounding the hole.

Other seventeenth-, eighteenth-, and nineteenth-century astronomers disagreed. Some thought the spots were not holes in clouds, but rather mountains on the Sun's surface, poking through a luminous liquid. Tobias Swinton, a clergyman, had an even more straightforward and perhaps maximally logical idea. In 1714, he wrote a book explaining that the Sun was simply hell, period. He argued that the popular notion of hell being located far below Earth's surface was obviously wrong, because any fire there would soon be extinguished by the lack of air, and hell has to keep burning. Plus, Earth's interior is too small to accommodate all the damned, especially when one makes allowances for future generations of the damned-to-be.

The idea taken most seriously was that the Sun got its heat and substance from comets that were continually pulled in by its tremendous gravity. More sophisticated analysis, however, made this hard to accept: given the Sun's great outflow of heat, the number of incoming comets would have to be crowded and continuous, and there was no observational evidence for this.

One refinement, by mathematician and theologian J. Wiedeberg in 1775, was that the Sun ejected countless minuscule particles that we experience as heat. When these particles stuck together, they formed spots, which hovered immediately over the solar surface. He went on to say that some of these particles were pushed away from the Sun at high speeds by "solar electrical forces"; when far enough away, they coalesced to form planets and comets. This is how Earth was created. Our world is a sunspot.

Although the church could no longer dictate what scientists wrote, many authors nonetheless tried strenuously to make any hypothesis compatible with the Bible. Of particular difficulty was Genesis, which tells us that God created light on the first day, but the Sun not until the fourth, later even than the creation of Earth and grassy meadows. In other words, our world and its green fields, nicely illuminated, endured for several Bible days without the Sun. It's a tough thing to explain, but that didn't stop many from trying.

The award for most creative might go to Charles Palmer, who, despite being an accountant rather than an astronomer, published a widely read theory in 1798 saying that the Sun is made of ice. Yes, ice. Since light (says the Bible) existed before the Sun, the Sun could not be the source of light. Therefore, the Sun must only reflect and concentrate the universe's light and focus it on Earth, like a magnifying glass. Indeed, isn't the Sun round, just like a convex lens? And don't the stars and the Milky Way have their own light, independent of the Sun? Clearly, the Sun is not the supplier of light, but merely the instrument for getting it to Earth. But what could this giant lens be made of? Not glass; that would be man-made. It must be ice.

Palmer offered further proof that the Sun is not the source of light and heat. Everyone knows, he said, that you feel more heat from a hot object the closer you get to it. But if you climb a mountain and get closer to the Sun, you feel *less* heat. Bingo.

Which of these views prevailed? Controversy and debate were the rule, of course, but schoolchildren during the nine human generations between 1650 and 1880 were taught that the Sun had a cool, dark interior and was covered with glowing, molten liquid. In addition, they were informed of a different, minority view, which also maintained a cool interior for the Sun but had it extend upward to a cool surface, surrounded by those fiery clouds.

As for the true source of all this salutary light and warmth, the texts of the time nearly always pleaded ignorance, beyond hazarding one or two vague and unconvincing guesses. Authors of the period commonly adopted a refreshing humility that these days seems to be in short supply when we hypothesize about things such as the birth of the cosmos.

Here's a typical description of the Sun from an 1847 text, *Celestial Scenery,* published by E. C. & J. Biddle of Philadelphia: "In the present state of our limited powers, we can form no mental image or representation of an object so stupendous and sublime [as the Sun]. Chained down to our terrestrial mansion, we are deprived of a sufficient range of prospect, so as to form a substratum to our thoughts, when we attempt to form conceptions of such amazing magnitude."

The humble, flowery preamble dispensed with, the author then serves the meat: "From a variety of observations, it is now pretty well determined that the solar spots are depressions, and not elevations, and that the black nucleus of every spot is the opaque body of the Sun seen through an opening in the luminous atmosphere with which it is environed."

And that is where solar knowledge stood for 250 years. Then a series of unexpected scientific breakthroughs once again animated the wheels of progress. The information these new scientists uncovered was more than amazing. It lay beyond anyone's dreams for one simple reason: it focused on the invisible.

CHAPTER 5

The Unit

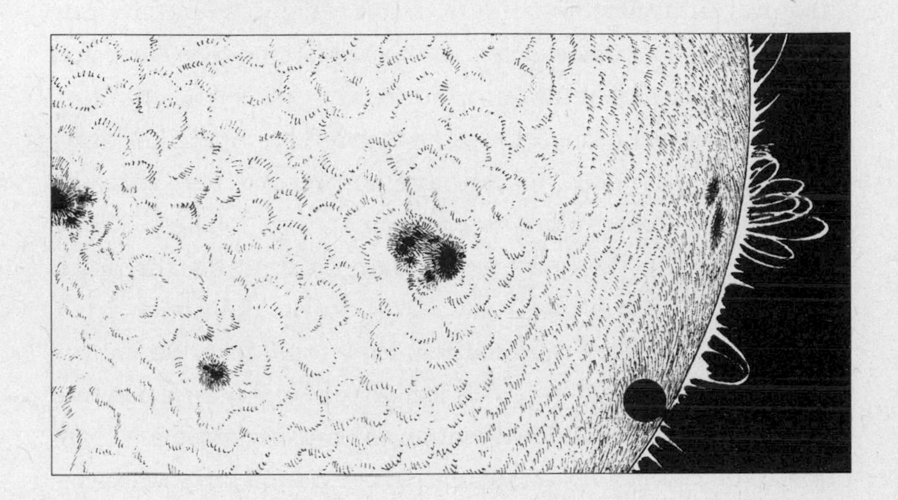

Here comes the sun, here comes the sun,
and I say it's all right.

—George Harrison, "Here Comes the Sun," 1969

THE EIGHTEENTH CENTURY witnessed an amazing event. It was astonishing, actually. The war-prone, pre-Victorian world put aside its enormous cultural and language differences to scientifically unite in a single epic and perilous multinational pursuit.

Our present twenty-first-century generation cooperates globally when we build the International Space Station or share advancements in cancer therapy. Such costly efforts, spurred by altruistic and intellectual rather than financial motives, represent the very best of what it means to be human. But the first such global undertaking, which unfolded 250 years ago, had a solitary fanatical focus. You will not be surprised that it centered on the Sun.

Of all the questions asked about that nearest star, none was more basic than this: how far is it? This was a concept to which everyone could relate in that time of steadily improving cartography. The issue endured even while sunspot fascinations came and went. Ignorance about the Sun's distance from Earth was a collective embarrassment that lingered on through lifetimes and wasn't solved until the 1960s. But far more was at stake here than the Sun's distance alone. Thanks to a proclamation by Johannes Kepler in 1619—his third law—we would learn the exact distance to all the planets the moment we knew the distance to the Sun.

This simple decree was amazing. How could it be? Surely, the Sun was merely a single thread in the tapestry of space. Yet Kepler's careful observations and brilliant mind correctly convinced him that the square of any planet's time period around the Sun exactly equals the cube of its distance from the Sun. Wait, hang in there. This is too amazing not to enjoy, and it requires the investment of only one minute.

We do not easily know the distance to planets—how could we? But we can certainly jot down how long it takes them to circle the Sun. We see it happening. We watch each planet chugging like a train through the zodiac constellations. After a bit less than 12 years, Jupiter completes one circuit around our sky, which also means one trip around the Sun. We observe Saturn lugubriously orbiting the Sun once every 29½ years. Speedy Venus does it in 225 days.

Let's stick with Jupiter. According to Kepler (who had thought intensely about orbital geometry and math for more than a decade, after poring over detailed records of where each planet was observed to be on a nightly basis), if we square Jupiter's 11.86-year orbital period (11.86 × 11.86), the result, 140.6, should be exactly the same as the cube of its unknown distance from the Sun. So all we need to do is find the number that, when multiplied by itself twice over, equals 140.6. This "cube root" of 140.6 turns out to be 5.2 (meaning 5.2 × 5.2 × 5.2 = 140.6). Voilà! Jupiter's distance from the Sun must be 5.2.

But 5.2 *what?* What distance unit does this equation employ? Kepler decided that one unit would equal the distance from Earth to the Sun. Indeed, astronomers everywhere still call the Sun–Earth span "one astronomical unit," or 1 AU.

So by merely observing Jupiter's track against the stars, which had been carefully watched and noted by various cultures since even before Moses said his first word, people now knew, thanks to Kepler, that Jupiter's distance from the Sun is 5.2 AU. And since Earth, by definition, is 1 AU from the Sun, Jupiter's distance from us must be its own distance from the Sun, 5.2 AU, minus Earth's distance from the Sun, 1.0 AU, or 4.2 AU.

This was one of the greatest revelations of all time. Overnight, the world, with some quick math, knew that Saturn must be 9.5 AU from the Sun and thus 8.5 AU from Earth. The whole solar system was now laid out. Scale models were built and were all the rage. You could place a big yellow Sun ball in a garden and a little blue Earth ball 10 feet away. Mercury's correct place was then 4 feet from the Sun, Venus 7 feet, Mars (outside Earth's orbit) 15 feet, and then, after a huge gap, Jupiter at 52 feet and Saturn at 95 feet. People did this everywhere, especially at fancy colleges such as Eton. Some teachers today still set up such models on school playgrounds.

There was just one problem. Sure, we now knew that Venus is 0.7 AU, or 70 percent of our own distance, from the Sun. But what in tarnation did that come to in miles? See, Kepler's third law gives the proportion, the scale, and the model. But he had no way of attaching true distances to it. It would be like Marco Polo triumphantly announcing that he had found the distance from Athens to Karachi, but he could express it only as "halfway to Beijing." That was useful, but people would still ask whether the trip entailed traveling one hundred miles or ten thousand.

As the seventeenth century clicked over to the eighteenth, educated people knew that if they could somehow figure out the true mileage to the Sun — the value of the astronomical unit — or find the distance to any planet whatsoever, they could then fill in the rest of the model, and they'd know it all.

But how? No tape measure was long enough.

What about parallax? Close one eye, hold up a finger, and notice what part of your distant wall it's in front of. Then close that eye and open the other one — in other words, alternately blink each eye — and the finger's position jumps. (I remember this very activity occupying much schooltime as an alternative to whatever was happening at the blackboard.) That is parallax — seeing things from different angles. If you measure how much the finger shifts, you can trigonometrically calculate how far your finger must be from your eyes. This method was now tried with the moon. Observers in two cities a known distance apart would note what background stars were next to the moon at the same prearranged time, then compare notes. The amount the moon shifted was all they needed to know. Joseph Lalande successfully nailed down the moon's distance by this method in 1751. It was pretty easy, since the moon can reposition by twice its width as seen from widely separated locales.

Some people tried to do this with the Sun. But its brightness even during an eclipse made the process far more difficult, plus the shift was super-tiny because of the Sun's much greater distance from Earth. Observers also tried to determine the distance of planets such as Mars, but the measurements were tough. The crude results did yield some rough idea of the whole scale, however, and by the early eighteenth century, the astronomical unit seemed to be somewhere between 85 million and 100 million miles. Not bad.

But knowing the exact value became a holy grail. A milestone was reached in 1716 when the British superstar astronomer Edmund Halley publicized the fact that pinning down the distance to any planet would yield the Sun–Earth distance, the astronomical unit, via Kepler's third law. He showed that the nearer the planet was to Earth, the better, because its parallax would then be larger (your finger shifts more if it's closer to your blinking eyes), and hence any error would be less important.

Looking at Kepler's scale-model distances, the nearest planet must be Venus, just 0.3 AU from Earth, with Mars in second place at 0.5 AU. We now know that Venus can approach to flirt, at just 26 million miles away. That's still a hundred times farther from us than the moon. So Venus doesn't switch positions in the sky very much when observed from different cities. Its "jump"—even when viewed from maximally far-apart places on Earth—amounts to the apparent width of a quarter-dollar coin seen from half a mile away. Not easy to measure.

Worse, Venus is extremely hard to see when it's nearest to Earth, because it's then almost in line with the Sun. But there's one loophole: even though its orbit tilts with respect to Earth's, once in a blue moon Venus passes directly in front of the Sun. It's then a fairly large, round, black dot crossing the Sun's face, an event called a *transit*. What if two observers far from each other—one in Beijing and one in Brooklyn, say—noted the moment when they saw Venus touch the inner edge of the Sun's disk at the conclusion of the transit? Comparing the times this happened at each place, knowing the distance between the observers, and knowing Venus's apparent speed as it crosses the Sun's face, they could then obtain the parallax and calculate Venus's distance with accuracy. Bingo: the value of the astronomical unit and the completion of the entire solar system jigsaw puzzle.

Psyched, Halley said that "the Sun's parallax may be discovered to within its five-hundredth part" if the transit could be "obtained true within 2 seconds of time." Optimistically—too optimistically, it turned out—Halley believed that such precision lay within the capability of the existing "telescopes and good common clocks."

Astronomers of the time knew that Venus's transits are rare. No, "rare" isn't quite the right word; it's more like "bizarre." Once a transit happens, another will follow in exactly 8 years minus 2 days. But then 105½ years must pass, followed by another pair of transits 8 years apart. Then 121½ years. Then 8. So that's the curious pattern: 8, 105½, 8, 121½, 8, 105½, 8, 121½, and on and on forever. And the transits are always in December or June.

I clearly remember as a ten-year-old looking ahead to the great celestial events of the distant future, when life on Earth would be so different and everyone would have a personal robot and flying car. There was Halley's comet in 1986 (Edmund Halley is always a role model for geeky ten-year-olds), then the mythical far-off turn of the millennium in 2000 (or 2001, depending on how accurate you wanted to be), and then, finally, the Venus transits. There had been none—not one—in the entire twentieth century, which fell within the longest possible transitless gap, 121½ years. Indeed, only six transits of Venus have ever been recorded. They took place in 1639, 1761, 1769, 1874, 1882, and, most recently, on June 8, 2004.

You don't even need a telescope to see one. Just a solar filter so you won't turn your retina to charcoal. In 2004, the sky was cloudless at my home in the mountains of upstate New York. And there it was, a respectable black dot moving leisurely across the Sun's face. It would be just eight years until the next, on June 5, 2012, and as in 2004, the entire United States would be on the correct side of the spinning Earth to see it, or at least some of it. Venus takes hours to strut across the solar disk, and the fact that the Sun will set in North America before Venus completely finishes its transit hardly matters. A few hours' worth of a transit is more than enough.

You won't want to miss the next transit. Get hold of eclipse glasses or shade No. 12 welding goggles, and look at the Sun late in the afternoon on June 5, 2012. Miss it, and you'll have to wait 105½ years for the Christmas transit of 2117.

Back in 1716, Halley's "Observe the next transit!" cry became a worldwide call to action. A fellow Brit, the future astronomer royal George Airy, announced a century later that finding the astronomical unit was "the noblest problem in astronomy." It made the popular press and got everyone aroused in an era when, granted, there wasn't much on TV.

Only a single human being had ever knowingly observed a transit. That had been, Halley pointed out with pride, a British cleric named Jeremiah Horrocks. He was only twenty at the time, the age of a college junior today. He had been born into a common family but was brilliant in math, and by strength of mind he had managed to squirm his way into Cambridge. Without the aid of any of our modern computing aids or tables, or even calculus, which had not yet been invented, Horrocks figured out that Venus would cross the Sun's face on November 24, 1639. His math was detailed enough for him to know that the transit would be visible from England. The real miracle, of course, is that when the day arrived, the weather was clear in Lancashire. Horrocks observed the event and recorded its details for posterity. He even had a witness, having convinced his friend William Crabtree to look for it, too, in Broughton.

After the transit, Horrocks offered an estimate of the Sun's distance. Aware of Kepler's third law, published just twenty years earlier—in the year he had been born—Horrocks realized a full lifetime ahead of Halley how Venus's distance could yield the Sun's distance. But, alas, there were no other observers far enough away with whom he could compare notes to obtain the parallax. Using indirect methods alone, Horrocks found that the astronomical unit was 59 million miles. This is, of course, far from the correct distance, 93 million miles, but it was the best calculation to date.

Sadly, Horrocks died suddenly and without any obvious cause at the age of twenty-two, in January 1641. William Crabtree, for the rest of his life the only person on Earth to have witnessed a transit, said of his friend, "What an incalculable loss!" Today essentially unknown, he is immortalized by the lunar crater Horrocks.

Toasting Horrocks, who had died seventy-five years earlier, Halley threw down the gauntlet in 1716, eleven years after he predicted the arrival of the eponymous comet that would ultimately arrive to bring him immortality after his death. Although he projected a sense of urgency, there was truly no need to rush. The next transit was still forty-five years in the future. The long countdown began, and astronomers and governments responded by raising impressive funds for expeditions to be dispatched with the best clocks and the best telescopes for the long-awaited transit of June 6, 1761. It was a matter of national pride and prestige to find this near-biblical grail.

Eight European nations sent out expeditions. At least 120 observers, including the top astronomers of the day, sailed, hiked, rode horseback, and trudged to sixty-two locations in a world that was only 14 percent as populated as today. Astronomers traveled as far as India, South Africa, Constantinople, Beijing, and Siberia. These observations were so important that the French and British—in the midst of the fierce and deadly Seven Years' War, fought even in their territorial possessions around the globe—officially provided each other's astronomers safe passage. It was a costly international effort, the likes of which the world had never seen. And it was all in the service of obtaining a single number.

The far-flung expeditions encountered epic adventures, trials, and even death. They faced dispiriting problems using equipment in steamy humid environs, not to mention overcast skies in other places. Consider the travails of the French astronomer Guillaume-Joseph-Hyacinthe-Jean-Baptiste Le Gentil de la Galaziere, who tried to observe the transit in India. He failed to arrive in time, thanks to an outbreak of war (and probably delays caused by officials writing his name on passport applications). Appalled, he decided to remain on the subcontinent to catch the next transit, which would take place eight years later. Alas, when those long dysentery-plagued years finally elapsed, he saw nothing because the skies were overcast thanks to an early monsoon. A dispirited Le Gentil returned to France, only to find that his heirs had assumed he was dead and had taken his property.

When the dust settled and all the weary observers had returned from their five dozen destinations, published papers started to appear. However, as more and more of them went to press, it became clear that the astronomers had failed to achieve what Halley had hoped. The parallax angle obtained for Venus varied by nearly 25 percent. Halley had expected accuracy of one part in five hundred. This was more like one part in five. With Venus at these widely derived distances, the corresponding Sun–Earth span ranged from 81 million to 98 million miles. In fact, the astronomical unit was now fuzzier than before the transit. The observers might as well have stayed home.

But they didn't. The same nations now rolled up their sleeves for their next (and final) transit opportunity, which would come in just eight years.

Actually, the 1761 transit produced some intriguing if weird reports. Several astronomers saw a ring of light around Venus when it was almost in front of the Sun and concluded—absolutely correctly—that the Sun, from behind, must be illuminating a thick Venusian atmosphere extending fifty miles into space. But many teams had trouble pinpointing the all-important exact moment when Venus was within and tangent to the Sun's disk because of a bizarre effect called the *black drop.*

Instead of a dark, inky Venus in cameo, sitting at the zero hour with its curved edge just perfectly and momentarily meeting the Sun's inner edge, a black protrusion like a ligament strangely joined the two for several seconds. It resembled what one might expect if the blackness of space just beyond the Sun's edge was a kind of ink that leaked inward to meet the ebony Venusian disk as it approached the solar limb. No one could figure it out, but it threw off the timing.

Today we know that this is a diffraction effect caused by the interference between light waves. You can duplicate it this very moment if you hold up your thumb and forefinger in front of a bright wall or the sky. Held half a foot from your eyes, slowly bring your thumb closer and closer to your finger, as if about to make the "okay" sign. Move your thumb extremely slowly, and you'll see that just before it makes contact with your finger, a strange black blob fills in the small empty space. This is what appeared between Venus and the Sun's edge, and it messed up everything. (See the black drop illustration on p. 51.)

Preparations were even grander for the transit of June 3, 1769. Instead of eight European nations, there were eleven. More than 150 astronomers and assistants, and a small army of support staff, set sail to seventy-seven locations around the world. Everyone involved in this stupendous collaboration and competition, even boatswains and kitchen staff, knew that they sought one solitary correct number. It was all in the name of science, with no practical benefit whatsoever, no hope of it yielding some new medicine or technology or vast riches.

Nearly half the ships were sent out by Britain alone, one of them led by Captain James Cook. For his ship, however, the transit was a cover. His was an intriguing spy mission. He was given the secret assignment of trying to learn whether Australia ("the new southern continent") actually existed or was merely a myth. While he was at it, he was to explore the little-known islands of the South Pacific.

On August 26, 1768, Cook's *Endeavour* set sail from Plymouth with a crew of ninety-four and with eighteen months' worth of supplies. This was Cook's first major voyage, and he was not fooling around. After rounding the always perilous Cape Horn, with its characteristically wild seas, Cook reached and anchored at Tahiti at the beginning of June 1769, just in time for the transit.

Hearing that Cook had left a monument behind, I traveled to Tahiti (someone has to do these things) in 1986 and found it on a lonely hilltop outside Papeete, at a place named Point Venus in honor of the expedition. The barren, windswept promontory was deserted. The plaque was understated: "This memorial was built by Captain James Cook to commemorate the observation of the Transit of Venus, June 3rd, 1769."

Cook's notes for the day read as follows:

> Saturday 3rd This day prov'd as favourable to our purpose as we could wish, not a Clowd was to be seen the Whole day and the Air was perfectly clear, so that we had every advantage we could desire in Observing the whole of the passage of the Planet Venus over the Suns disk: we very distinctly saw an Atmosphere or dusky shade round the body of the Planet which very much disturbed the times of the contacts particularly the two internal ones.

> Dr Solander observed as well as Mr Green and my self, and we differ'd from one another in observing the times of the Contacts much more than could be expected.

All the expeditions were better prepared than in 1761. The telescopes were bigger and finer, and the clocks were as well. And the results? Not perfect, but better. The Venus parallax results ranged from 8.3 to 8.8 arc seconds, which led to a Sun–Earth value of somewhere between 93 million and a little more than 97 million miles. Astronomers decided that the best average was 95 million, and this was the astronomical unit taught to schoolchildren for the next century.

No one alive for the 1769 transit was still around for the next pair in 1874 and 1882, and a new group of astronomers were now raring to try their own luck. Eighty expeditions were dispatched around the world, and once again the transit was a big hit in the popular press. John Philip Sousa even wrote the "Transit of Venus March" in 1883 to celebrate the event, although this was not one of his big hits. Photography had now been invented—the first picture of the Sun had been taken in 1845—and the annoying black drop could be photographed rather than just observed visually. Still, the measurements were even better, and the astronomical unit was refined to 92,880,000 miles, a mere 76,000 miles from the actual value. Yet such an error would still mean that a spacecraft launched from Earth today would totally miss its intended planet.

It took until 1958 for "the noblest problem in astronomy" to be nearly laid to rest. That year, Paul Green and Robert Price, two Massachusetts Institute of Technology (MIT) engineers, sent the first radar pulses to Venus, achieving results that even they could not duplicate and thus were not scientifically acceptable. Three years later, in March 1961, Richard Goldstein of Caltech nailed it.

Timing the radar pulses and knowing the speed of light allowed Goldstein to define the astronomical unit and fill in the entire solar system once and for all—just in time for the burgeoning space age.

What is this number, this fundamental cornerstone for expressing all planetary distances? It is 92,955,807 miles, Earth's average distance from the Sun. It would have been lovely if it had been a perfectly round number, such as 100 million, but that would have been uncanny (and strong evidence that everything is a dream).

Edmund Halley would have traded his comet for that number. Yet tell it to your friends, and odds are they will shrug. Astronomy is, of course, a heavily numbers-oriented science, but the public is overwhelmingly unaware of such basics as the Sun's size, the Earth's diameter, the galaxy's width, or the distance to the moon. Still, an informal survey I conducted twenty years ago showed that the rough distance to the Sun is one of only two celestial figures that many people know. The other? Virtually everyone can recite how many planets are in the solar system, although this has been thrown into recent disarray with Pluto's demotion.

In January 2000, on the newly minted prime-time TV show *Who Wants to Be a Millionaire,* contestant Dan Blonsky reached the final question: "The Earth is approximately how many miles away from the Sun?" He had four rounded-off answers from which to choose: 9.3 million, 39 million, 93 million, and 193 million. He was moments away from being financially set for life. The audience sat, tensely silent. Only one other contestant in game show history had ever won that much money.

Blonsky's eyes went from one choice to another and back again.

Captain Cook would have slapped his head.

CHAPTER 6

Magnetic Attraction

Busy old fool, unruly sun,
Why dost thou thus,
Through windows and through curtains, call on us?
—John Donne, "The Sun Rising"

AFTER 1720, SUNSPOTS streamed back over the Sun's face like a swarm of locusts. Their bizarre near-total absence was soon forgotten. Since a common phenomenon generally arouses less interest than a rare one, and since spots were still regarded as mere curiosities, their increased numbers led scientists to emit a loud collective yawn and turn to other matters, such as the Sun's elusive distance.

Into this period of extended ennui entered Heinrich Schwabe, born in Germany in 1789. Bored with his profession as an apothecary, he turned his focus to the real love of his life, astronomy, happily making the leap from elixirs to nebulae. Schwabe began to observe sunspots in 1826. Actually, he didn't care very much about the spots themselves, which were probably as tedious to fixate on as saltpeter. Rather, he was trying to discover a new planet.

Just forty-five years earlier, William Herschel had become the greatest astronomer in the Milky Way after he found Uranus. Overnight, his name was known by just about everyone. To duplicate that feat was every nineteenth-century science enthusiast's dream.

But Schwabe made a big mistake. Instead of looking farther out to where the next planet, Neptune, was destined to be found just twenty years later, he searched within the orbit of Mercury, seeking a scorched inner world many believed must exist. It had even been given a name: Vulcan.

Charbroiled Mercury whirls closely around the Sun like a moth. It never stands farther from the Sun than 27 degrees, little more than the span between an outstretched pinkie and thumb held up against the sky, which prevents it from ever appearing against a dark night backdrop. Even the great Copernicus supposedly never saw Mercury. How much more difficult to spot, then, must Vulcan be? It would surely never leave the Sun's glare. Schwabe believed that one promising way to detect the planet would be to glimpse it as a dark spot passing in front of the solar disk. This reasoning was sound. After all, we've already seen how Venus conspicuously transits across the Sun's face twice a century, while Mercury performs the same feat thirteen times in the same interval. Vulcan, which must possess an even shorter and faster orbital path, should logically appear in front of the Sun quite frequently. Its discovery, it would seem, required only that someone dedicate all his waking hours to continuous Sun staring.

On every non-overcast day for seventeen years, from late 1826 to early 1844, Schwabe observed the Sun and recorded its spots, hoping to find an imposter—one that would appear perfectly round and move relative to the others. He was so diligent that he managed to find enough desperate breaks in the clouds to make drawings of the Sun three hundred days a year, which is almost supernaturally successful in cloudy northern Europe. Of course, he didn't find Vulcan, which didn't exist. But seventeen years into his project, he noticed a very odd thing. For five years or so, the number of spots kept increasing until they reached a maximum, and then they fell off for a nearly equal period of time. Somehow, no one had ever noticed this before. Schwabe made this pattern a central point when he wrote an article with the soporific title "Solar Observations During 1843," and it was here that the world first learned of a sunspot period.

Schwabe's paper was widely ignored. But a lightbulb went on in the brain of a Swiss reader named Rudolf Wolf, who happened to be the director of the Bern Observatory. Wolf, one of those high-metabolism people with boundless energy, immediately set out to confirm this solar rhythm with his own sunspot observations.

At around the same time, the charismatic German explorer and naturalist Alexander von Humboldt—whose exploits, writings, and fame were spreading throughout Europe—wrote a massive bestselling book covering every imaginable science subject. In this work, *Kosmos,* he cited Schwabe's recent findings about the ten-year sunspot period, and this is how it became widely known. It was highly effective publicity. It would be as if some nineteenth-century Jay Leno started his monologue with this joke: "This guy named Schwabe gets pulled over by a constable and says his carriage was on the wrong side of the road because he keeps staring at spots on the Sun. (*Audience laughs.*) Claims they come and go every ten years, like his marriages. (*Audience howls.*) And guess what? Turns out this guy is really a druggist!"

The blockbuster *Kosmos* led to Schwabe's payback for those seventeen tireless years of Sun watching. He was awarded the prestigious Gold Medal of the Royal Astronomical Society in 1857. Now, a century and a half later, most researchers still call the Sun's famous eleven-year sunspot period the *Schwabe cycle.*

Heartwarming, sure, but by itself this would have merely introduced yet another inconsequential solar oddity. The critical second part of the story raises the curtain on a new and unexpected phenomenon: magnetism.

These days magnetism, to most people, is something odd and minor. In grade school, most teachers place a piece of paper over a horseshoe magnet and sprinkle iron filings on it. The particles land in a strange, alien-looking curved pattern that the teacher may call "field lines."

Field lines? Curved streams of influence that force each bit of metal to end up here but not there? It's an impressive and thoroughly bizarre demonstration to the third-grade brain. Yet who among us imagined it had anything to do with the sunshine that hit our cheeks when the school day was over? Even as adults, few of us are aware that light itself is a self-propagating wave that alternates between being a pulse of electricity and a pulse of magnetism.

Magnetism was well-known to the ancient Greeks. Aristotle said that Thales had discussed it at length three centuries earlier, around 600 BC. At the same time, early Dravidian records mention an Indian surgeon named Sushruta, who supposedly used a powerful magnet to remove iron debris from wounds.

Magnetism was also familiar in ancient China. A book from the fourth century BC casually notes, "The lodestone makes iron come to it." Back then, even Chinese peasants knew how to use tiny bits of lodestone to keep from getting lost. Called a *south pointer,* a magnetic splinter was gently dropped onto the surface of water, where it would whirl around and magically align itself with the north–south direction. It took until the 1200s for Europeans to begin doing the same thing, but then they soon started mounting the magnet in a dry housing, where the needle would swivel on a point.

In 1600, just as poor Bruno was being burned at the stake, the English physician William Gilbert had his groundbreaking book *On the Magnet, Magnetic Bodies, and the Great Magnet of the Earth* published. Happily for him, it offended no one. Gilbert was the first to conclude in writing that Earth itself is magnetic, and this is why compasses point north. This was huge. It seems obvious to us today, but throughout history people confronted with natural oddities tended to be either too myopic or too farsighted and searched for causes either immediately nearby or else in some unimaginably distant realm among the stars. Hence, many thought that an undiscovered magnetic island in the icy northern sea was pulling on compass needles, while others imagined the attraction came from the North Star itself.

By the time compasses became commonplace in the West in the thirteenth century, it was clear they had several quirks. As a pilot and airplane owner, I'm trained to be wary and aware of these vagaries, which can easily throw off one's navigation by a dozen degrees. (Sadly in a way, modern GPS equipment has made the kerosene-filled wet compass virtually obsolete, even though all planes still have them. Pilots tend not even to glance at them until they attract attention to themselves by leaking and filling the cockpit with frighteningly volatile fumes, which they dutifully do every few years.) Among the eccentricities of compasses is that they do not point to the north and south poles of Earth's rotation. If they did, compass needles would show true north, and life would have been much easier for navigators through the centuries. Instead, compass needles align themselves with field lines, which culminate at our planet's *magnetic poles.*

Earth's north magnetic pole, where field lines dive vertically into the ground and toward which compasses point, currently sits five hundred miles from the geographic pole around which we rotate. Specifically, it's just west of Ellesmere Island in northern Canada, where nobody in his right mind has ever lived. This spot was first located by explorer James Clark Ross on June 1, 1831. There's no buoy or marker or Starbucks there, because Earth's magnetic poles shift, and they've been moving faster and faster of late. During just the past century, the magnetic north pole has migrated a whopping 650 miles; in the past forty years, the rate has accelerated from 5 miles to 37 miles a year. That's 16 feet per hour. You could barely have dinner at the magnetic pole before it goes somewhere else. Undoubtedly, the cause is sloshing liquid iron in Earth's outer core.

Keeping track of these changes is easy, because compass needles essentially point to the new polar spot wherever it is. Since it isn't true north, navigators must apply a correction, called *deviation,* to figure out where actual geographic north lies. In Boston today, all compasses point 15 degrees west of true north.

One fun fact, rarely realized, is that uncorrected compasses point toward auroras. In the eastern United States and Canada, that's 10 to 20 degrees left of true north. On the west coasts of both countries, it's 10 to 20 degrees *right* of north.

Some places in the world happen to sit in the correct position so that magnetic north lies in the same exact direction as true north, and many maps include a dotted "agonic line" showing this ribbon of zero compass deviation. In the United States, that line runs from Lake Superior south to the Gulf of Mexico.

Another baffling magnetic vagary — fully realized by the early nineteenth century — is that compass needles point to the right of magnetic north in the morning and to the left in the afternoon. Compounding this oddity, the amount of this variation changes and is rarely the same two weeks in a row.

Fortunately, magnetism, whose appeal as a respected and well-funded research subject ebbs and flows, was hot during the first half of the nineteenth century, stirred by the German genius mathematician and astronomer Carl Gauss. At about the same time, explorer Alexander von Humboldt spent five years sailing around South America, keeping obsessive records of the local magnetic field and its changes wherever he went. In 1852, the German astronomer Johann von Lamont, poring over decades of magnetic records around the world, including Humboldt's, realized that the daily variations in the compass displayed a pattern: they grew larger for five years, then smaller again, in a cycle spanning just over ten years. He had already written about magnetism in his 1849 *Handbuch des Erdmagnetismus* (Handbook of Terrestrial Magnetism), and he found this extra twist strange and interesting—but then thought no more about it.

Remember Schwabe's 1843 discovery that sunspot numbers wax and wane in a cycle of about ten years, and—bringing things full circle—how this was publicized in Humboldt's *Kosmos*? Now the stage was set, and all that remained was someone to put two and two together. Three Europeans who were avid readers of the latest science—Rudolf Wolf, Edward Sabine, and Alfred Gautier—independently connected the dots. They all realized at virtually the same moment that there existed separate published knowledge that the sunspot cycle and Earth's magnetism shift in unison in matching periods. As Wolf explained in 1861, "This correspondence between two phenomena, one of which seemed until now to belong exclusively to the Sun, and the other exclusively to the Earth, was extraordinarily remarkable."

Remarkable, indeed, that tiny spots on the Sun physically moved the needle encased in the fancy engraved instrument found in nearly everyone's home. But how? What was the mechanism?

Wolf got to work. First, he continued his assiduous sunspot

observations started a few years earlier. As director of one of Europe's most prestigious observatories, he had no trouble corresponding with colleagues around the continent. All agreed to record sunspots in a manner that would make the data standard and comparable. Then he went to libraries and searched for every possible record of sunspot observations he could find, dating all the way back to Scheiner and Galileo 250 years earlier. In the end, Wolf secured records for 22,500 days of sunspot observing, the equivalent of sixty-one unbroken years. He not only obsessively counted the spots but also started weighting the data to compensate for the varying techniques of different observers. One can only imagine his exasperated wife: "Rudy! Dinner's been ready for five hours. The venison is charcoal. Why must you count sunspots every waking moment? Is it me?" Wolf's likely reply: "7,133 . . . 7,134."

Being a mathematician by training, he applied valid statistical methods and concluded that the time between one period of maximum sunspot activity and the next is just over eleven years. He even realized that the cycle varies from eight to seventeen years and that the interval from minimum to maximum is shorter than the other "half" of the cycle. To keep track of these epic breaths of the Sun, he numbered them, starting with the 1755–1766 cycle. That was "one." We are now in cycle 24, which began in 2008.

When Wolf published this exhaustive body of work, many scientists remained skeptical. He had to wait thirty-three years, until three more complete solar cycles had passed and everything still fit, before his work was universally accepted. It didn't hurt that as an increasingly popular author of books on astronomy and other topics, he was gaining a growing readership. He also projected total confidence in his observations and analyses. Fortunately, he lived long enough to see his critics fall silent. Today astronomers still use what they call the Wolf sunspot number to compare similar sunspot groupings.

Wolf's life ended before the twentieth century, when magnetism finally coughed up its most fascinating secrets. Only then could we determine the historic direction of Earth's magnetic poles over millions of years, thanks to the discovery that iron in volcanic lava aligns with our planet's magnetic field as it solidifies (as soon as it cools below its "Curie temperature" of 768°C, or 1,414°F). And only then did we realize that our planet's magnetism is mysteriously weakening. It has decreased by 10 percent since Wolf's heyday. No one knows whether this is part of a larger, more profound alteration in our world's magnetism, or whether such fluctuations have always been a capricious, if benign, physical aspect of our planet.

The other part of the equation for Wolf—the *variation* in Earth's magnetic field, and thus how compasses behave on a day-to-day basis—was much trickier to confirm historically. Accurate magnetic variation measurements extended back only a few decades. There was no Galileo or Scheiner taking compass readings 250 years earlier to allow for long-term comparisons. Based solely on the few cycles he had in hand, Wolf predicted Earth's magnetic variation for the year 1859, and his prediction proved to be spot-on.

But more than that, everyone wanted to know how sunspots could move earthly objects. Indeed, maybe things happened the other way around. For all we knew, perhaps it was Earth's changing magnetism that caused sunspots! Three centuries earlier, the church would have adored this possibility, for it would have supported the old beloved "Earth is the center of the universe" paradigm. Or maybe an outside factor affected both the Sun and Earth simultaneously. For a while, Wolf thought that perhaps the giant planet Jupiter influenced the rest of the solar system. After all,

Jupiter's orbital period of 11.86 years was temptingly close to the average sunspot cycle. Had Wolf known what we know today—that Jupiter has the strongest magnetic field of any object for eight and a half light-years in every direction all the way to the Dog Star, Sirius—he would surely have never abandoned his Jupiter idea, but instead erroneously run with it. As things stood, however, the lack of any known mechanism linking sunspots with Earth's magnetism just hung in the Great Hall of Science without a caption, as enigmatic as women's fashions in Uzbekistan.

The twin eleven-year periods were indeed new and exciting, but it was not an entirely novel notion that magnetism was somehow mixed up with spots. A century earlier, the French naturalist Jean-Jacques Dortous de Mairan, after assembling historical observations of European displays of the northern lights, realized that, like sunspots, they were absent between 1620 and 1710. He correctly suggested that the two phenomena were linked. Of course, he had no idea how and lamely guessed that the atmospheres of the Sun and Earth merge from time to time.

Edmund Halley did better. After witnessing an amazing eight-hour auroral display over London in March 1716, he rightly conjectured that it might have some connection with magnetism. He never guessed that the northern lights involve the magnetism of Earth and the Sun together, and that either of them alone cannot do the job. Instead, he wrote, "[Magnetism does not] otherwise discover itself but by its effects on the magnetic needle, wholly imperceptible, and at other times invisible, [but] may now and then by the concourse of several causes very rarely coincident, and to us yet unknown, be capable of producing a small degree of light."

In short, Halley thought that magnetism is normally an invisible phenomenon, like gravity or love, but that once in a while, it glows as the northern lights. This shows how Halley could really think outside the box. The idea that magnetism sometimes materializes visually was a very cool concept, and we know now that when it bends the paths of electrons, the result is an eerie blue light called *synchrotron radiation,* which lights up the Crab Nebula. He also thought that maybe an intermediate step was involved, during which magnetism draws out something from within Earth, and it is this that fluoresces as the aurora.

So great minds were trying, and thinking, and one or two linked the aurora with sunspots. A few others connected the aurora with terrestrial magnetism. But nobody correctly fashioned a tapestry woven of all three threads. To do so, they would have had to flash forward to the 1980s, since picking the minds of even mid-twentieth-century scientists would not have done the trick. Even when the first astronauts were hopping around on the moon like kangaroos, astronomers were still baffled by how the Sun's corona, its atmosphere, could sizzle at 2 million degrees Fahrenheit when its surface below was just a lukewarm 11,000 degrees. It didn't make sense: where was the oven? Only in the 1980s were the answers laid bare—and they brought magnetism once again to the front page.

It turns out that sunspots are places where magnetism is much stronger, which blocks the motion of material rising up from the Sun's interior. Slower-moving matter is cooler. Cooler means darker. And that, very simply, is a sunspot.

Above the Sun's bright visible surface (the photosphere) where magnetic field lines flow in giant arcs, sometimes the magnetic lines disconnect and snap open. The sudden break releases energy and whips particles that were following these paths to superfast speeds. Fast means hot. Bingo.

So all the Sun's visible phenomena come from magnetic variations, violent interplays, disconnections of field lines, and reconnections. In turn, these complex interplays stem from a bizarre, newly discovered boundary layer inside the Sun, 70 percent of the way from the core to the surface. The vast region inside of this, the innermost Sun, is called the radiative zone and rotates uniformly as if it were a solid ball. The region outside it, the convective zone, has that turbulent differential rotation, taking twenty-five days to spin at the equator but more than a week longer at the poles. The thin boundary separating these major parts of the Sun's body is called the *tachocline* and was discovered only in 1989. It is here that all the magnetic chaos and violence erupting from the surface is born.

As for its magnetic rhythms, it turns out that the Sun's heartbeats are sometimes more like arrhythmias. The strange, superlong sunspot minimum that persisted from 2008 through 2009 had its origin in the "conveyor belt" by which magnetized solar plasma flows along the Sun's surface toward its poles, where it sinks and returns toward the equator. Normally, these fiery rivers dive down and head back to the equator when they reach 60 degrees latitude, the solar equivalent of Oslo or Stockholm. But in our most recent solar cycle, the flows traveled all the way to the poles, and the return was slower, too, making the Sun more tranquil than anyone alive today has ever seen it.

Even the direction of its magnetic poles resists any desire we might have for solar stability. While Earth's magnetic field irregularly reverses its north–south polarity a few times each million years, the Sun, being gas and more dynamic, flips its poles for each new human generation — every twenty-two years or so.

That this magnetism, of all the crazy minor players, is the key to the Sun's visible changes was just starting to be suspected in the mid-nineteenth century. As seemingly disparate phenomena on Earth and its nearest star started coming together, speculation began anew: if the Sun and its spots were physically moving our compasses hour by hour, what else was the Sun doing to us?

No one suspected that the violent side of solar magnetism would cause billion-dollar destruction and take human lives in the century to come. Rather, there was a sense in the early nineteenth century—optimistically suspected more than logically surmised—that the world was on the verge of a revelation, that the Sun could not keep its fiery secrets forever. And, indeed, its true nature was destined to be exposed sooner rather than later.

CHAPTER 7

The Wild Science of the Bearded Men

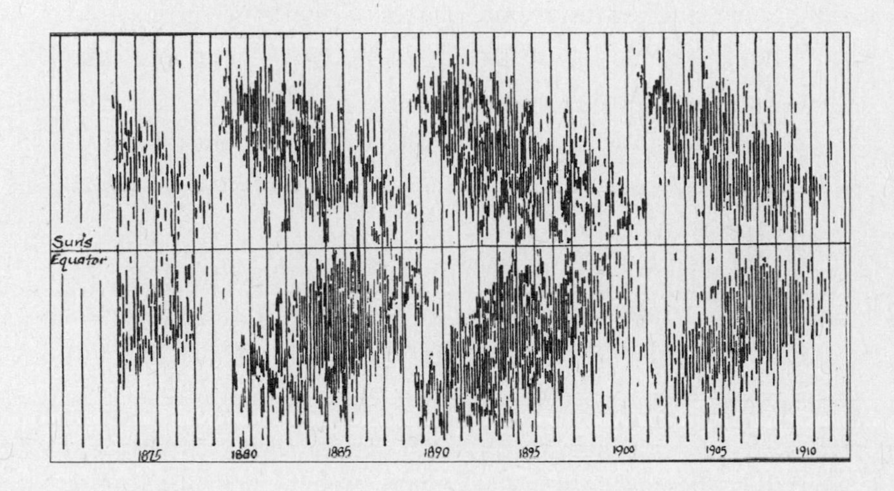

We need not hesitate to admit that the Sun is richly stored with inhabitants.

—William Herschel, 1818

THE SUN WAS still a cipher when the nineteenth century began. By the time it ended, much of our modern knowledge was in hand. This radical metamorphosis came courtesy of a small coterie of brilliant, obsessive oddballs.

But things looked bleak in the early 1800s. The Sun's Wikipedia entry had not changed much in two centuries, and neither, incidentally, had the world's physical understanding of Earth, which was still widely believed to be only six thousand years old.

Essentially, people knew enough to be depressed. Understanding the distance to the Sun, and especially to the stars, and the hopelessness of ever obtaining a sample for analysis, the popular French philosopher Auguste Comte used the Sun and stars as models of futility, of what humankind could never learn. In 1835 he wrote, "All investigations which are not ultimately reducible to simple visual observations are... necessarily denied to us.... We shall never be able by any means to study [the Sun and stars'] chemical composition."

Yet he was soon proved wrong. The unexpected breakthroughs started with a wispy, almost dreamlike observation that initially seemed meaningless.

William Herschel, already famous for finding Uranus in 1781, discovered infrared energy ("calorific rays") in 1800 by letting sunlight hit a prism and cast its rainbow spectrum on a thermometer (as we'll see in chapter 10). Suddenly, prisms were hot, and scientists were doing everything they could think of with them short of feeding them to goats. Two years after Herschel's infrared announcement, the English chemist William H. Wollaston let sunlight first pass through a narrow slit in a thin piece of metal before it entered a prism, in order to project the solar rainbow. This arrangement changed the spectrum in a very odd way. Now imposed on the vivid colors were several thin, black vertical lines, which Wollaston unpersuasively described in 1804 as "divisions between the colors," as if the Sun's spectrum was a kind of coloring book that needed outlines. Even he probably didn't believe this.

A decade later, the Bavarian optician Joseph von Fraunhofer created a better setup with a series of superior prisms and a narrower slit—an instrument that forever after became known as a spectroscope. The improved solar spectrum featured dozens of unvarying black lines, both broad and narrow. Fraunhofer, as baffled as Wollaston had been, did the only thing he could: he named the lines. The darkest and most prominent got letters of the alphabet starting from the red end of the spectrum. For example, a closely spaced pair of dark lines dominated the yellow region of the Sun's rainbow spectrum, and he called these D_1 and D_2. Today, if you speak with a physicist or high school science teacher, she'll know exactly what you mean if you allude to the Sun's "double-D Fraunhofer lines."

For the next half century, science books hungry for anything to say about the Sun mentioned that its light is crossed by dozens of dark lines of unknown origin and meaning. In other words, the Sun's seemingly white light is made up of colors with narrow gaps or absences, where, for example, a precise shade of blue-green or deep yellow is missing. This seemingly subtle phenomenon was destined to utterly change astronomy, chemistry, and physics, even though the lines remained a mystery for two more generations, until a pair of professors at the ultraprestigious University of Heidelberg, Robert Bunsen and Gustav Kirchhoff, entered the story in 1859.

Bunsen was already widely acclaimed as a chemist, having made breakthroughs in isolating the salts of arsenic (even though he is mainly known today for improving the laboratory gas burner so that it emits a flame so pure and hot it's virtually invisible). A few years earlier, a lab explosion of possibly the most stinking, foul substance on Earth, cacodyl arsenic, had sent shards of glass flying and cost Bunsen his right eye. This failed to dampen his enthusiasm, and the accident only added to his near-mythic renown as the world's expert on arsenic and its many faces.

His colleague and close friend Kirchhoff was even more famous, by far. Kirchhoff was a math whiz and one of the world's pioneers in understanding electricity; "Kirchhoff's laws," which deal with electrical circuits, are still taught today. Despite his own disability, which required him to use crutches or a wheelchair for his entire adult life, Kirchhoff was fun-loving. The renowned American physicist Robert von Helmholtz described him in an article in *Popular Science* as having "refined speech and a courteous and attractive demeanor. His sense of humor and his wit won him the liking of all men with whom he came in contact. His friendship with Bunsen [has] become very close. They take walks in company, and they travel together during vacations."

In 1859, the colleagues used Bunsen's superhot burner to vaporize elements until they glowed, then observed the light through a spectroscope. What they saw astonished them.

When liquid mercury (for example) was vaporized, the hot gas gave off a blue glow. However, when this light was passed through a slit and then a prism, it was seen to emit three vivid colors: violet, green, and yellow. No blue at all. The professors realized that the seeming blue of excited mercury is merely what we perceive when its actual spectral emissions are scrambled together in our eyes. In other words, a spectroscope displays the true colors given off by any heated substance. And, best of all, the pattern emitted by any particular element or chemical never varies.

When Bunsen and Kirchhoff dropped a pinch of salt into a flame (try it yourself), it produced a yellow flash. Through a spectroscope, it was clear that this light from sodium came from two nearly identical shades of yellow that formed a tightly spaced pair of golden lines.

So each element emits its own unique fingerprint. Kirchhoff and Bunsen realized they'd found a convenient new way to analyze substances: simply make the substance glow, look at it spectroscopically, and consult the chart they'd prepared listing each element's pattern. It was easy. It seemed too good to be true.

One night, the bearded buddies saw through their window that a distant house was aflame. They grabbed their spectroscope to look at the burning building and could see brief flashes of various colored lines as lead or copper pipes became hot enough to contribute to the inferno's light. Why, they learned, one can identify substances even from great distances!

In late 1860, they turned to the brightest fire of all, the Sun, and those inscrutable Fraunhofer lines. I wish I could have been present for the eureka moment. Bunsen wrote to a friend in England: "At present Kirchhoff and I are engaged in an investigation that doesn't let us sleep. Kirchhoff has made a wonderful, entirely unexpected discovery in finding the cause of the dark lines in the solar spectrum, and he can... produce them [in our lab] and in the same position as the corresponding Fraunhofer lines. Thus a means has been found to determine the composition of the Sun and the fixed stars."

Kirchhoff explained it himself: "The dark [Fraunhofer] lines... exist in consequence of the presence... of those substances which in the spectrum of a flame produce bright lines in the same place."

By studying the countless black-line patterns on the Sun, the two scientists matched them with bright-line patterns of dozens of elements—hydrogen, oxygen, iron, calcium. Here was the Sun's composition as surely as if it had been brought to the lab.

The key to gaining astronomical knowledge instantly changed, too. Merely looking at anything through a telescope was now demoted to a hobbyist's pastime of limited usefulness, since stars remain mere dots through even the biggest instruments. By contrast, if a spectroscope is placed where the star's pointlike image comes into focus, a gorgeous, colorful carnival of lines materialize to reveal the star's makeup. And its temperature. And its rotation, since spectral lines broaden if the star is spinning. And any unseen companions, since, in about half the stars, astronomers observe two sets of spectral lines. And its speed and motion in space, since the lines all shift toward the red end of the spectrum in a star rushing away from Earth and toward the blue end if it's approaching.

In double stars, one set of lines shows a redshift and the other shows a blueshift, then the pattern reverses itself over time. As a result, we know the period in which the two stars orbit around each other and thus take turns approaching or receding from us.

The wonders of the spectroscope go on and on. A spectrograph tells volumes, while a telescope shows only a point of light. No wonder astronomers since then spend 75 percent of their time and resources on spectroscopy alone. The popular image of a professional astronomer looking through a telescope is, sadly in a way, as archaic and incorrect as a savory airline meal in coach.

The importance of Kirchhoff and Bunsen's discovery cannot be overstated. It arguably remains the greatest scientific breakthrough of all time. And yet who knows Kirchhoff's name today, 150 years later? As for Bunsen, he is vaguely known, but only because of that laboratory gas burner he did not even invent but merely improved, still ubiquitous in high school chemistry labs.

In 1860, the world abruptly knew that the Sun is composed of the identical substances as Earth—in gaseous form, of course. This was actually no surprise. The most popular theories had said that the Sun and Earth formed together from a nebula, and if this was true, they should naturally be made of the same stuff.

By the late 1860s, all but a few of the solar Fraunhofer lines had been matched with earthly elements. The lone unidentified pattern gained more attention with each passing month, until it became as significant and irksome as an errant husband's late nights out. By 1868, physicists essentially gave up and acknowledged that this series of lines in the yellow part of the spectrum matched nothing on Earth. Here, apparently, was one substance the Sun possessed but Earth did not: a uniquely solar element. The notion held an elegant appeal. It was soon named *helium,* after the Greek Sun god, Helios.

Helium had its fifteen minutes of fame—actually a bit more

than that. Its status as the only exclusive Sun element lasted for two decades. Then the Scottish chemist William Ramsay, who discovered more elements than anyone before or since (and who today knows *his* name?) found helium in pitchblende mines, where it accumulates as a stable end product when uranium decays into lead. Ramsay rushed this newfound gas to his lab, made it glow, and bingo: the pattern precisely fit every remaining unknown line in the Sun's spectrum.

The spectrum story has one last key player, a young Dutch physics teacher named Pieter Zeeman. In 1896, disobeying orders not to use the school's lab equipment on his own, he discovered that a strong magnetic field makes spectral lines split into two copies, like a drunk seeing double. His reward for this unauthorized but world-class discovery? He was fired by his supervisor.

Zeeman got his revenge without even trying. He received universal acclaim, was offered a far more prestigious and vastly higher-paying post in Amsterdam, and won the 1902 Nobel Prize for what forever became known as the *Zeeman effect.*

This had several immediate consequences. First, his original mentor, Hendrik Lorentz, figured out that this splitting of lines must mean that the rapid motion of a tiny, negatively charged particle has to be the source of all light everywhere in the universe. This astoundingly sound logic took place three years before the actual discovery of the first subatomic particle—the electron, in 1899—whose movements are, in fact, the sole source of light. And, oh yes, Lorentz won the Nobel Prize, too, sharing it with Zeeman in 1902.

Second, and more to the point of our story, astronomers quickly isolated and observed the spectral lines from sunspots and found that they are indeed broadened, or split into copies. This means that sunspots are regions of strong magnetism. We know today that their magnetic field strength is three thousand times greater than Earth's. This was the first critical evidence for magnetism on the Sun and began the increasing recognition that magnetism is the key player in the dramatic solar phenomena that influence our world.

❧

SO FAR WE'VE MET a quartet of big nineteenth-century brains that immensely improved our Sun knowledge: the three bearded Germans, Fraunhofer, Kirchhoff, and Bunsen, plus the mustachioed Dutchman, Zeeman. I've saved the financially poorest for last. To meet him, we must cross the Channel.

Edward Walter Maunder was born, the youngest of four boys, in Victorian London in 1851 to a Wesleyan minister. A severe illness in his teens, from which he fully recovered, gave his mind the unexpected leisure to focus on science and nature observations, and thus he was observing sunspots when he was fourteen, while other boys were illegally downloading movies or whatever occupied teens of that era. E. Walter Maunder, as he started calling himself—simply Walter to his family and friends—then studied chemistry, math, and physics at a local college.

This was still the heyday of the British Empire and of scientific competition, particularly with Germany, so resources went to the sciences, especially the fields of navigation and astronomy. Being from a poorer family, Maunder could not attend a prestigious school; he supported himself with a full-time bank job, where his math skills were utilized and appreciated.

Fortunately for him and for us, his life radically changed when he decided to take a civil service exam in 1872, the first ever held in Britain, the result of liberal breezes that were starting to ripple across its rigid class system. It was a rare opportunity for a bit of upward mobility, but the odds were slim. The Royal Observatory at Greenwich had exactly one opening for an assistantship, and applicants didn't need a college degree to land the job. When the information was published in the newspapers, the flood of applicants caused the observatory's director to decide that maybe he could use *three* new assistants. Maunder's score was good enough to win him the third post.

The job started largely as unpleasant drudge work under the authoritarian directorship of George Airy. Still, he was fortunate: the observatory's copious daily arithmetic was performed by "computers" — male teenagers hired to do no more than perform calculations. These low-paid mental menials were kept on for a few years in factory-like conditions, working twelve-hour days as if they were stitching corsets, and then invariably discharged by the time they were twenty-one, with no hope of advancement. Maunder stayed on, however, and by 1875 he had saved enough money to marry.

The British had pioneered the use of "watch stations" throughout their empire. The people manning these stations took careful, frequent measurements of the weather, the current magnetic variation, and the appearance of the Sun and sunspots, partly for scientific prestige but mostly with the hoped-for practical advantage of predicting weather conditions for the empire's vast military and commercial fleet. Maunder's work mostly involved the Sun, although his specific job of "photographic and spectroscopic assistant" was a brand-new one. A main focus was sunspot observation, which was now, like the sunspot cycle itself, enjoying one of its "up" periods in its endless swings in science fashion.

The Royal Observatory had decent equipment that kept getting better: a spectroscope, a twelve-inch refractor telescope, and a colossal twenty-eight-inch refractor for photography. In 1881, William Christie took over as head of the observatory, and solar work became the top priority. Especially important was measuring the redshifts and blueshifts of spectral lines and how they revealed which parts of the Sun were rising up toward the observer and which were sinking back.

No one yet realized that the Sun's surface continually moves up and down in complex rhythms, like that of a woofer. It took until the 1960s for astronomers to detect and measure these sound waves, which led to an important new tool in solar science called *helioseismology.* (Just pronouncing this correctly produces satisfaction.) Today helioseismologists utilize these physical pulsations, detected mostly by spacecraft, to understand what is happening out of sight below the photosphere, just as earthquakes unlock secrets beneath our planet's surface.

At this point in his career, his beard still dark, the forty-four-year-old Maunder, whose wife had died in 1888, was raising his four surviving children on his own. It was now that he fell in love with that rarest of all rare creatures, the sole female computer working at the observatory. The brilliant Irishwoman Annie Scott Dill Russel — educated at Cambridge University's Girton College and having received top honors as "mathematician of the year" — would later become a world-renowned expert on solar photography. When she and Walter were married in 1895, she was forced by law to leave her job as a computer (which had no future anyway), although she continued to assist Walter professionally.

THE RESULTS OF Maunder's relentless sunspot observations from 1874 to 1902, plus his keen conclusions, aided by his and Annie's mathematical analyses, were finally published in the *Monthly Notices* of the Royal Astronomical Society in 1903, in a second piece in *Monthly Notices,* and in an article in *Popular Astronomy* in 1905 titled "The Solar Origin of Terrestrial Magnetic Disturbances." Here at last were the results of thirty years' worth of analyzing measurements on nine thousand photographs, of plots of the solar latitudes of some five thousand sunspot groups—the kind of tedious work that might have driven another man into the loony bin.

The peerless professional work created by the collaboration of Walter and Annie included the now famous "butterfly diagram," a term the two coined. (See Alan McKnight's illustration replicating it on the opening page of this chapter.) Back then, the quickest glance at the diagram confirmed Heinrich Schwabe's discovery of an eleven-year sunspot cycle a half century earlier. But it also revealed how the spots symmetrically change their location and number in both solar hemispheres. It showed that they are virtually nonexistent at solar latitudes north or south of 35 degrees, how they symmetrically migrate closer to the solar equator, and how the time of minimum spots is soon accompanied by the vestiges of a new cycle that manifests as maverick high-latitude spots both north and south.

Moreover, Maunder said that the spots are magnetic and influence terrestrial compass needles as well as the aurora borealis. He showed in particular how when unusually large spots reach the exact middle of the Sun plus two days' more of travel, terrestrial disturbances and auroral displays break out all over Earth. He demonstrated that sometimes these storms rotate off the edge of the Sun, reappear two weeks later, and again cause terrestrial disturbances soon after they reach the Sun's midpoint.

Maunder also documented a long historic period of sunspot and aurora absence from 1645 to 1715 and further noted the extremely cold weather during that time. Along these lines, he tried to publicize the very recent work of the German astronomer Gustav Spörer, who was the first to become aware of the dearth of seventeenth-century spots and also uncovered an earlier sunspot absence in the fifteenth century. Unfortunately, nobody seemed interested in Spörer or his pronouncements, and they stayed uninterested after Maunder cited him in his articles.

Were Maunder's hard-won conclusions accepted with applause and recognition? Not at all, even though the famed mid-twentieth-century solar researcher Jack Eddy named that amazing sunspot blackout period the *Maunder minimum,* a term now universally accepted. (He also called the earlier fifteenth-century blackout the *Spörer minimum,* but this bit of arcanum is probably too obscure even for Final Jeopardy!) You see, Maunder had a nemesis. And as bad luck would have it, it was Lord Kelvin, a man of popularity, achievement, and tenacity.

That intensely brilliant Scotsman, born William Thomson in 1824, had produced groundbreaking mid-nineteenth-century contributions to thermodynamics and electricity. He was also the first to put forth the concept of absolute zero, the utter absence of heat, and the world honored his insights with the Kelvin temperature scale, which is universally used in the sciences today. (It is the only scale logical enough to have its zero at the lowest possible temperature where all atomic motion comes to a frozen halt: −459.67°F, or −273.15°C.)

Thomson was respected by scientists and academicians but unknown to the public until he got involved with the design and installation of the first transatlantic telegraph, starting in 1854. All of Britain was intrigued by the seemingly impossible job of laying three thousand miles of continuous undersea copper cable along the perilous ocean floor. Adding to the drama, Thomson had ongoing disagreements with the man in charge. They fought over important fundamentals, such as what diameter cable was necessary, how the signal should be boosted, and what voltage was needed to send at least one coded telegraph character every four seconds. Specifications called for the capability of transmitting fifteen dots and dashes—a single five-letter word—each minute. (A word a minute? These days it may seem amazing that the world could ever have applauded such a glacial communications rate. Even if it was successful, a man utilizing it could consummate a marriage faster than he could propose one.)

Thomson's boss always got his way, and in the end he messed things up big-time. After two costly attempts failed, the "successful" third attempt failed as well—fried by improper voltage. Thomson's ongoing concerns were proved justified. He was now put in charge, and he got the job done. For his success and the swelling English pride, he attracted much newspaper attention and was knighted. He took the name Lord Kelvin.

As brilliant as he was, Kelvin became dotty in his old age and began making weird pronouncements. During the final decade of the nineteenth century, just as Maunder was completing his work, Kelvin announced that X-rays were a hoax and didn't really exist. He also expressed the strong opinion that airplanes were impossible. And when he applied his mind to the big solar problems of the day, he made a fundamental mistake.

He stated publicly that Maunder's conclusions had to be wrong, because sunspots could not possibly send so much power to Earth as to disturb its magnetic field or cause auroras. He published figures to prove it. He imagined that the Sun must directly radiate magnetism the way it radiates light, which he said would then inversely drop off in intensity with the cube of distance and, moreover, become diluted to oblivion by spreading like light in all directions. Earth would receive only a tiny fraction of this energy, never enough to move a compass needle. Something within Earth must be responsible for the aurora and all the other disturbances, he insisted. He was sure of it. He wrote, "It is absolutely conclusive against the supposition that terrestrial magnetic storms are due to magnetic action of the Sun.... The supposed connection between magnetic storms and sunspots is unreal. The seeming agreements between the periods has been mere coincidence."

Kelvin was influential, and his views were parroted by textbook authors of the time. Against all this stood Walter Maunder—the commoner, the nobody—and, unseen behind the curtain, Annie.

How could terrestrial magnetic disturbances *not* be from the Sun? argued Maunder to the few who would listen. Compass needles everywhere hugely deviate when a giant sunspot reaches a point 17 degrees west of the Sun's midpoint, he pointed out, and then go crazy all over again twenty-seven days later when that same sunspot rotates around to reach 17 degrees west of the solar meridian once more.

Maunder could not rebut Kelvin's mathematical calculations, because they were flawless. But what he now insisted in print was that Kelvin's *premises* had been wrong: "The origin of our magnetic storms *does* lie in the Sun. It is not from the whole surface; it is not radiated equally in all directions."

Walter and Annie later rhetorically asked, "Can sunspots, faculae, prominences and changes in the corona affect us here on Earth even if we hid ourselves underground, shut off by many feet of cold ground from the sight of the Sun, from the knowledge of day and night, or summer heat and winter cold? The answer is yes. Even if a magnet is suspended deep in a cellar, far from the madding crowd, it will quiver."

Maunder even had the correct reason for this. It is not magnetism itself that the Sun throws our way. Rather, when a sunspot is in the right position, it uses its magnetism to hurl "corporeal particles" toward Earth. Here was the world's first description of a solar wind.

In 1908, Walter and Annie coauthored *The Heavens and Their Story*, in which they eerily summarized terrestrial magnetic effects this way:

> [A magnet's needle] does not remain undisturbed for long. From about nine in the morning till about two in the afternoon there is a feeble swing of the magnet to the west, during the remaining hours it creeps back. Day by day the magnetic needles swing to and fro, but the extent of the swing is not always the same.... The swing is greater in the summer months, when the Sun is high above the horizon, than in the winter months.
>
> Every now and then... a monster spot breaks out upon the Sun. It passes across the Sun's disc [and], without warning, the magnetic needle, swinging gently to and fro in the stillness, becomes violently agitated. It quivers as if it were transmitting a panic-stricken message. Who sent this message? What does it mean?

They even thought that we can see the "message" being broadcast. Annie's photograph of the 1898 solar eclipse in India shows long coronal streamers, which look like rays heading out to space. What if one of these was aimed right at our planet?

Kelvin's arguments to the contrary were trumpeted by a few people for a while, but not for very long by Kelvin himself. Failing health robbed the Scotsman of his fire for confrontation, and he died in 1907.

At the same time all of this was transpiring, Maunder was attracting personal followers and fellowship. In 1890, he founded the British Astronomical Association, with an initial membership of 283. A sort of support group, it was an association "of amateur astronomers, astronomers for mutual help, who because of their sex or other circumstances might be precluded from joining the Royal Astronomical Society." Maunder's organization was an antidote to the Lord Kelvins of the world, and maybe even some payback on behalf of all those who were then treated unfairly by the established sciences. Above all, the British Astronomical Association was inspired by Annie, who despite passing her final qualifying exam at Cambridge and being voted mathematician of the year, was not awarded a degree solely because she was a woman.

For Annie and other turn-of-the-century women of science, the Sun was finally rising. Not so for the couple's groundbreaking discoveries. Maunder's findings, including the amazing fact that the sunspot cycle had stopped cold 250 years earlier, were quickly forgotten. They were not published in any encyclopedia or science textbook, but instead languished in oblivion until resurrected and popularized by Jack Eddy in the 1970s.

Walter Maunder spent his final years writing and editing in association with such prestigious journals as *Nature.* He passed away in 1928, at the age of seventy-seven. Annie went on to become a noted astronomy book author and chronicler of solar eclipses. She died in 1947, at age seventy-nine. On the far side of the moon and on the planet Mars, craters named Maunder now honor them both.

CHAPTER 8

Cautionary Tales

Sol *gold is, and* Luna *silver we threpe,*
Mars yron, Mercurie *quiksilver we clepe.*

— Geoffrey Chaucer, "The Canon's Yeoman's Tale"

EVEN EIGHTEENTH-CENTURY schoolchildren knew that Sol rules day and night, summer and winter, life and death. In contrast with today's widespread obliviousness, the lowliest tradesman of 250 years ago knew the details of the Sun's annual patterns of solstices and such. What was new was the freakish realization that visible changes on the Sun somehow physically reach down and move the compass needle in everyone's home.

What else might it do? A bit ominously, the Sun was increasingly suspected of holding hidden power over human lives. Perhaps the Egyptians had been right to fear Ra.

Marks on the Sun had been linked with earthly consequences even in the time of the poet Virgil (70–19 BC), who wrote, "When [the Sun is] checkered with spots in the early dawn, beware of showers." The earlier writer Theophrastus (ca. 372–287 BC) similarly connected rain with the sighting of sunspots. Nearly two thousand years later, the Italian Jesuit Giovanni Riccioli, who called the moon's dark blotches "seas" and inconsistently named them for either emotions or weather, as in *Sea of Tranquility* and *Ocean of Storms,* wrote in 1651 that "an abundance of spots" is associated with "cold and rainy weather." This belief was widely repeated then and as much as a century later.

William Herschel, after discovering Uranus, decided to investigate this link. By comparing the recorded month-to-month price of wheat in Windsor, England, with the number of sunspots for sixty-three years starting in 1650, he found that a scarcity of spots predicted when wheat would be more costly. More modern studies have, incidentally, shown the same thing. (Some researchers still think this connection holds water, but I've never seen BLTs get cheaper during solar maxima or noticed a "sunspot special" on any menu.)

When Herschel published this sunspot–wheat price connection in 1801, few took it seriously. Opinion on the street remained split as to whether spots meant cool weather or the opposite. In reality, seeing fewer sunspots *is* strongly linked with cooler earthly weather (we recently saw this dramatically from 2006 to 2009 — an observation co-opted by skeptics of global warming), although Herschel's wheat price studies occurred when the solar cycle was essentially gone, making those particular year-to-year changes unrepresentative.

We must, at some point, honestly confront a major obstacle in the enjoyable pastime of trying to link sunspot activity with earthly events. Today we have available records of only three dozen full solar cycles since the days of Galileo and Scheiner. We attempt to match up any of the thousands of terrestrial phenomena that tend to be cyclical with this small sample of solar ebbs and flows. Nothing stops us from seeking matches between times when spots are either scarce or common and times of war and peace, economic booms and busts, harvest yields in all countries and all crops, weather and storm anomalies, animal populations and declines, sexual mores and fashions, political parties in power—really, anything at all. With only thirty-six sunspot cycles to compare any of these events with, we will, of course, find many seeming correspondences. The law of averages insists on it. Indeed, we do find that the sunspot cycle ebbs and flows correspond with the length of women's skirts, the rabbit population of Australia, and the party that controls the US Congress. But are these things causally linked with sunspots, or are they coincidences?

This is one of the great differences between New Age reasoning and the application of scientific and statistical methods. In the former, there are no coincidences, because everything in the universe is linked with everything else. "It's all One," say some of my favorite starry-eyed friends. The problem, of course, is that even if this is true, it is of no help whatsoever in deciphering the specific workings of the cosmos.

In trying to rebut such logic and illustrate why any search for truth requires skepticism, I told one friend, "Suppose, just as you brushed your teeth this morning, a gust of wind knocked over an old white pine tree in Maine. The two events were perfectly simultaneous. Would you call this a coincidence or correspondence? Did your toothbrushing kill that tree?" Most people would concede that mere coincidence was at play.

But imagine if your grandmother died just as lightning struck the village church and stopped its clock. As hard-nosed and skeptical as we might be, wouldn't most of us link the two events together? Wouldn't we try to find some metaphysical meaning in the fact that the clock stopped ticking just as she took her last breath?

Researchers today are quite sure that the sunspot cycle does indeed correlate with Earth's temperature, the amount of solar energy it receives at all wavelengths, the appearance and frequency of auroras, the position and strength of the Gulf Stream, the thickness of the atmosphere, the health and longevity of Earth satellites, the clarity of radio transmissions, the corrosion of pipelines, the robustness of a few crops such as wheat, the amount of radiation we receive from the sky, the condition of the ozone layer, the amount of pre-cancerous damage our cells' DNA receives, and a great many other phenomena we'll discuss later.

But it cannot be stressed enough how careful we must be when we look for such links. The most obvious peril is that when we seek a connection between *any* pair of events, we will usually find it. This happens because when we're looking for correspondences, we become sensitized to those times the two things happen in unison and tend to overlook the times they don't.

An excellent case in point is the supposed link between human births and the full moon—which also involves the Sun, because every full moon happens when the Sun and moon sit on opposite sides of the sky like a seesaw. When one is rising, the other is setting. In reality, then, a full moon link with human births is also a Sun link.

I've found that almost everyone who works in the maternity wing of a hospital expresses the belief that births do march in tune with the full moon. "I've been here twenty years," one nurse told me emphatically, "and I don't just think so; I *know* so. We certainly do see more births at the full moon. Ask anybody."

She's right, in a way. Medical professionals commonly report more births at the full moon. It's fascinating—because it's wrong.

If births were cyclical in any way, birthdays would show a reliable sine wave pattern that, in fact, does not exist. So why the strong *perceived* relationship between births and the full moon?

Whenever a lot of births coincidentally occur during a full moon, hospital staff typically say, "It figures; there's a full moon tonight!" But when the full moon lands on a slow night, nobody says anything. There is no activity or event to spark discussion. Over time, the "more births at full moon" idea is reinforced, and even on hectic cloudy nights when the moon is neither full nor visible, some nurses still exclaim, "Must be a full moon!"

So how can we figure out what's what with either sunspots or the full moon? Only through statistical analysis. While not particularly difficult, statistics must be applied correctly, which is why a professional statistician is now routinely part of every science study, no matter the field.

Many investigations have looked for a link between births and the moon. Any association would be doubly interesting because the average length of a human pregnancy, 266 days, happens to match the 265.8 days of nine complete cycles of the moon's phases (synodic months). This means that any lunar connection with births is also a link with time of conception and thus fertility. Let me put it this way: on average, the moon's appearance, or phase, at one's birth is the same as it was at one's conception. And since it is the Sun that illuminates the moon at various angles to create those phases, the relative positions of both the Sun and the moon at conception are on average repeated on the day of birth.

It turns out that all lunar and solar configurations appear to have the same connection with fertility and birth: the connection is random. Any differences are either extremely small or nonexistent for the vast majority of women. We know this because numerous studies have yielded conflicting but generally negative results.

With all the controversy this topic creates, I think it's worth a closer look.

The most positive findings come from a 1959 study by the father-and-son physician team of Menaker and Menaker, who looked at 500,000 births in New York City public and private hospitals. They then performed a follow-up study of 501,000 births, published in 1967. The first investigation showed a slight (1 percent) rise in births in the three-day period around the full moon. The second study also showed a small increase, but this time just after the first quarter moon.

The ten-year period spanned by these studies and the great number of births involved make them the best argument for a weak link between births and the moon and Sun. Taken together, however, they do not support a connection with the full moon, but rather a 1 percent birth difference between the two halves of the lunar month. Specifically, they found an average of 455 births during the "bright" lunar half of each month and 451 births during the "dark" half. It's hard to see how such a subtle difference would be noticeable to any medical staff.

And the news only gets worse. A 1973 study of 500,000 additional births in New York City produced an even weaker link, again with the first quarter moon. A 1986 French study examining nearly 6 million births yielded further contradictions; it found more births at the third quarter and new moons, and fewer births around the first quarter and full moons. Entirely rebutting *any* lunar link were investigations at clinics and hospitals in Vancouver in 1975, in France in 1991, and in Florence, Italy, in 1994, as well as a number of studies in the United States, all of which showed no connection at all between date of birth and lunar phase, and hence also Sun position.

Even the studies containing tiny positive findings acknowledge that other factors almost surely affected the results. Several holidays, such as Easter, Passover, and some Muslim observances, automatically and invariably occur at the new or full moon, when the Sun and moon are together in the sky or exactly opposite each other. So is it the moon or Sun that influences human conception — or having a few days off from work?

Perhaps the most definitive evidence comes from a review of fifty years' worth of moon-birth studies published in 1989 in the journal *Psychological Reports.* After statistically evaluating twenty-one studies, the authors reported that there is no connection between human birth and lunar phase, period.

And yet medical practitioners continue to insist that such a link is strong and ongoing. The conclusion is obvious. We humans fool ourselves all the time. When looking for lunar or solar influence, we need to rely only on the analysis of trained statisticians. Moreover, one researcher may think she's seen enough to conclude that *x* causes *y*, while another is waiting for further proof. How much evidence is enough?

That's an important question when we get to such highly charged topics as solar warming of Earth versus anthropogenic climate forcing. Are *we* altering the planet, or can we finger our cellmate Sol?

Seeking causal links between phenomena became important when the scientific method gained favor in the seventeenth century. But as the late twentieth century brought a rise in global temperatures, a renewed caution against being too quick to perceive connections started to compete with a sense of urgency — and a growing call to action.

CHAPTER 9

Why Jack Loved Carbon

Neither the sun nor death can be looked at with a steady eye.

—François de La Rochefoucauld, *Maxims,* 1665

AFTER WALTER AND Annie Maunder died, their work died with them. Solar cycles and the Sun-Earth connection once more fell out of favor as a research subject. This is not to say that the first half of the twentieth century saw no solar advancements. To the contrary, this was when science finally uncovered the fusion mechanism by which the Sun shines. Nonetheless, the Sun-Earth temperature and climate link was kept alive by only a small handful of scientists.

You cannot blame mainstream scientists for their lack of interest. It's easy to find weather phenomena, especially regional or local ones, that appear to be in perfect sync with either the Sun's eleven-year cycle of spots or its twenty-two-year cycle of magnetic polarity.

For example, the water levels in Africa's Lake Victoria were found to match the eleven-year cycle, and this discovery, along with accompanying graphs, was published with great fanfare. No sooner did it appear in print than the two phenomena diverged, never again to repeat. Researchers were fooled over and over again by mountains of evidence of "matching" events that ultimately proved spurious.

Still, the Sun-Earth link was kept alive, if on life support. In the early twentieth century, Yale professor Ellsworth Huntington believed that periods of numerous sunspots correlated with storms and rainfall in enough places that it cooled Earth. He wrote, "Present variations of climate are connected with solar changes much more closely than has hitherto been supposed."

In 1919, Arizona astronomer Andrew Douglass published a list of connections between the sunspot cycle and tree rings. By personally measuring seventy-five thousand tree stumps, including extremely old sequoias (did he have time for anything else?) and the lumber used in North America's oldest buildings, Douglass concluded that the thickness of tree rings depends on solar variations. He also found that trees were dramatically different during the long seventeenth-century sunspot absence we now call the Maunder minimum. It took nearly half a century for other scientists to agree and widely value tree rings for climate study.

Through the 1930s, the main voice arguing for the Sun-Earth climate connection was Charles Greeley Abbot of the Smithsonian Astrophysical Observatory. He attempted to continually measure the "solar constant"—the intensity of sunlight falling on Earth—and announced that it showed large changes that he said were due to spots passing across the Sun's disk. He claimed that the long-term differences amounted to 1 percent and insisted that *of course* the sunspot cycle explained Earth's changing temperatures, once one factored in the temporary cooling spells caused by dust from volcanic eruptions. With the weight of his colleagues at the Smithsonian behind him, this view that sunspot variations were a main cause of climate change remained a nagging idea into the 1960s.

The problem was that it was easy to say such things but not so easy to come up with an actual mechanism by which the Sun accomplished climate change, especially when more accurate measurements of the solar constant showed that it really only rose and fell by 0.2 percent in visible light, albeit by at least 3 percent in ultraviolet and more than 15 percent in far UV. Was this enough? If so, it should allow predictions, but these kept failing. Eventually, science was thoroughly frustrated by dead-end sunspot-weather links. As one meteorologist recalled, "The subject of sunspots and weather relationships fell into disrepute. For a young [climate] researcher to entertain any statement of Sun-weather relationships was to brand oneself a crank." A big part of the problem was that the data was still unclear. Satellites had yet to be launched, and truly accurate observations of solar fluctuations (and, indeed, earthly global temperatures) were not yet in place.

By the late 1970s, it was crystal clear that carbon dioxide was increasing yearly and that temperature must logically respond. This got everyone's attention. Yet when the popular media tackled the topic of man-made climate change, they flitted from one idea to another. Back then, a couple of self-proclaimed experts announced that human soot and smoke were *cooling* the planet and that we were on the verge of a catastrophic artificial ice age. The fact that global temperatures did stay level for about thirty years, from the mid-1940s to the late 1970s, buttressed this view. But in all honesty, it was a fuzzy mess.

This is when Jack Eddy entered the scene. His influence on our current thinking still looms large. Indeed, he would become the most famous solar scientist in history. If you, dear reader, have never heard of him, that's merely proof that most superstars of science are destined to live and die in obscurity.

In the "most famous" categories of Jeopardy! or Trivial Pursuit, we'd all probably call out the same names. Go ahead: Who's the most famous sculptor of all time? The most popular singing group? The best-known rocket designer? Explorer? Early astronomer? Cellist? Baroque composer? There's a good chance you said or thought Michelangelo, the Beatles, Wernher von Braun, Magellan, Galileo, Yo-Yo Ma, and Bach.

Now try this: most famous chemist, biologist, physician, engineer, climatologist. No clear-cut winners. But if you worked in the physical sciences, you'd associate Jack Eddy with the Sun before you'd ever think of Galileo, Ra, Sun-Maid raisins, or anyone or anything else.

In my two decades with *Discover* and *Astronomy* magazines, I never interviewed Eddy, and by 2010 it was too late: he had died the previous June at age seventy-eight. I happened to mention this during one of my conversations with solar astronomer Spencer Weart, former director of the Center for History of Physics at the American Institute of Physics. "Listen," he said, "I interviewed Jack Eddy about ten years ago. He was fascinating. You can use the transcript if you'd like."

Much of the following comes from the transcript of that 1999 interview, in which Eddy recounts the heady discoveries of the 1950s, 1960s, and 1970s, when so many utterly new revelations materialized. In describing the notion that the Sun's output, or *solar constant,* changes over short periods of time with dramatic earthly consequences, Eddy said:

> The solar constant was to most solar physicists a joke. It was like saying, "I'm doing my thesis on the solar plexus. Or suntan lotion." It had so little to do with what they saw as the really exciting parts of the Sun. At the High Altitude Observatory...you'd work on long-standing but prosaic features such as flares or faculae or prominences. These other things that had to do with the Earth were considered a bit beneath an astronomer, and certainly an astrophysicist, and were best left to the amateurs, or maybe to someone at the fringe.

Among the top iconoclastic thinkers of that time was the American physicist Eugene Parker. "Gene Parker was an occasional visitor at the High Altitude Observatory in Boulder [Colorado]," Eddy recalled. "I remember him coming there in the late 1950s and giving a colloquium on his wild idea that there was a solar wind, before that was accepted by anyone. I remember the unanimous reaction of the solar physicists to his idea: what a ridiculous notion that was — a wind blowing charged particles out of the Sun."

Eddy continued:

> I had been taught that while the Sun indeed affects the upper and outer atmosphere of the Earth, purported connections with the troposphere and weather and climate were uniformly wacky and to be distrusted... for there is a hypnotism about cycles that seems to attract people. It draws all kinds of creatures out of the woodwork. One of those that turned up was this notion that Gene [Parker] told me about. About the work of Walter Maunder one hundred years before, when he had thought that there was a prolonged period of time in the 1600s when the Sun wasn't so active.

> That really piqued my curiosity, and I began digging into it. The trail was initially...driven by my prejudice of trying to find examples from the past that would disprove, once and for all, the notion of strong Sun-weather relations. A devout negativism on this subject was...the catechism that I had been taught and had taught to others.

Partly for his unorthodox views, Eddy was fired from his job at the High Altitude Observatory in 1973. A quarter century later, Eddy remembered that low point of his life:

> I was kind of desperate, for I had a family of four children—two in high school—to support. So I got a job to write this book. That kept income coming in. It also enabled me to continue work on the Maunder minimum, although not openly. It was a job that I had to stretch out as long as I could until I could get a [permanent] job. It became for me, intentionally or unintentionally, a kind of Scheherazade exercise, to be prolonged as long as possible.

It was at this point that a metamorphosis in Eddy's thinking occurred:

> I was beginning to change my mind. I knew I would face an uphill battle convincing my colleagues about the reality of the Maunder minimum if it leaned entirely on accounts from so long ago. Why should you trust someone in the 1600s when we are so much smarter and know so much more now? As scientists, we're trained to discount what one finds in old books, I think. Some things may be of interest for historical reasons, like whether Galileo was left-handed, but usually not for practical or applied ones.

> And I thought there ought to be some way to check on what Maunder, and earlier Spörer, had claimed.
>
> Because I had been trained in astrogeophysics and knew something of the other ways that the Sun affects the Earth, I looked hard at historical records of aurorae [and] Oriental naked-eye sunspot [observations] in the hundreds of years before the Maunder minimum and after.

In a way, Eddy was trying to retrace the steps of Gustav Spörer, who had done an amazing job combing through the European libraries of his day — practically living at some of them — before discovering what no one had even suspected: two separate periods of extended solar unconsciousness centuries earlier. Spörer had been ignored, as had his champion and popularizer Maunder. What made Eddy think he could add anything now, another century later, let alone sound enough alarms to get noticed?

But Eddy brought a new idea to the table: the Sun's atmosphere, its corona. He told Weart:

> I had gone to a bunch [of eclipses] by then, and what I thought was, well, if the sunspot numbers were really low, the corona should have looked a lot different. And then I tried to picture in my mind what the corona would look like if you really turned the sunspot number down almost to zero and kept it there for ten or thirty or forty years. What would the corona look like? It should be very dim, with few coronal streamers, if any. The real-time accounts seemed to confirm that.

Eddy said "seemed" for good reason. People tend to see what they are looking for, and eclipse observers in the fifteenth through nineteenth centuries were not focused on that seemingly inconsequential irregular glow surrounding the blacked-out Sun. Too much else was going on during totality, and the corona got little, if any, written mention in eclipse accounts, even by such articulate observers as Mark Twain, who took the trouble to travel to see one.

But Eddy noticed that during the Sun's supposed deep-sleep periods, no observer had *ever* reported a corona with the round shape it invariably displays during copious sunspot years.

> I also visited a bunch of museums, to look at how the Egyptians had portrayed winged Suns and whether the winged Sun wasn't an awful lot like how the corona appears at the minimum of the sunspot cycle. And it was, because it was all concentrated in the equatorial regions.
>
> When the work got published [in 1976] and became accepted, I was in this interesting position of having been canned by an observatory that wanted me back on board.
>
> I wanted to publish it in *Science* because I thought the story was of broader interest than to solar physics alone.... I thought, *They're going to go for this.* So I sent it there, and they did, and it landed on the cover. Part of the story also is where the name came from.

Like those working on Madison Avenue, Eddy had realized that a catchy phrase is all-important when offering a paradigm shift to a world ruled by inertia. He was determined to learn from the frustrations of Spörer and Maunder and to spin this dramatic slice of solar history into something a bit less geeky and easier to pass on to others. Eddy told Weart:

> Maunder was a kind of a second-tier astronomer at Greenwich, so far as I could ascertain. So when he started uncovering this stuff—much as I later uncovered what he had

> done—he didn't get that much press for it. I knew I had a lot of selling to do if people were to accept the notion of such irregularity in the Sun, and I sought a name that people would remember. "Maunder minimum," with all those *m*'s, had a kind of onomatopoeia.

After his Sun articles started appearing in popular publications such as *Natural History* and *Scientific American,* Eddy was halfway home: "When I read Maunder's and Spörer's claims from the 1890s that several hundred years before that, the Sun had behaved strangely for seventy years, it seemed almost preposterous. [But now] I believed it was a profound finding." Alluding to Maunder's history-making butterfly graph, Eddy said, "No one had ever seen that before, with this big flattened thing in there from 1645 to 1715, suppressing the normal eleven-year cycle."

Eddy, like most converts who give up eating meat or smoking cigarettes, shifted from teaching his students that solar cycles do not affect us to being an energetic, almost obnoxious proselytizer on radio and TV talk shows, as if to expunge his record of disseminating falsehoods. Now he had in his arsenal not just the historical records but a new weapon—an odd form of carbon that had been brought to his attention by none other than Eugene Parker, the discoverer of the solar wind. "The connection with carbon 14, I think, was the clinching thing," he recalled, after acknowledging Parker's crucial role.

> I had become certain enough on the basis of historical records, and aurorae, and the Chinese sunspot records, and what seemed to me the absence of the corona at that time. Even though none of these lines of evidence might have been strong enough to make the case—to convict a criminal in court—the combination of them all pointing

> the same way was to me more than convincing. It was like the strength that can be found in thread or string when enough strands are woven together. And then I looked into the carbon-14 thing and...it really seemed to fit the climate record....It looked like if you stuck in the solar key, it just hit all the tumblers in the climate lock. It was really something.
>
> The Sun really does go through prolonged periods of anomalous behavior, and will again someday.

After Eddy, nothing was ever the same, and the notion that global changes are connected with the Sun's variability remains powerful to this day. But what has really changed since Eddy's heyday is increased reliance on that amazing tool Eddy alluded to—carbon 14. You might assume that it's some kind of geeky thing, but I'm going to prove how cool carbon 14 is. First, however, bear with me a minute while we explore how it's made. And for this, let's review how atoms work.

Atoms have substantial, chewy centers made of protons and neutrons stuck together by the most powerful force in the universe, which, in the great poetic tradition of physics, is officially called the *strong force.* The number of protons in an atom's nucleus is super-important. This alone determines what element it is. This alone gives it all its general properties. If an atom has six protons, it's carbon. If it has just one more proton, it's nitrogen and doesn't resemble carbon in any way.

All atoms except the most common kind, hydrogen, also have neutrons in their nuclei, and these are just along for the ride. They don't alter an element's basic properties, and you'd feel equally comfortable breathing oxygen atoms with eight, nine, or ten neutrons. Neutrons don't help us and don't hurt us; they're like coleslaw. Each element has a normal number of neutrons, but there are "off-brand varieties" with one or two extras, called *isotopes.* They're specified by a subscript number added after the element's name. For example, ordinary oxygen's nucleus has eight protons and eight neutrons for a total of sixteen nucleons, so it's called O_{16} — or, because it's so common, simply O. But a tiny percentage of the oxygen in our bodies has one or two extra neutrons, and these isotopes are called O_{17} and O_{18}.

Ever wonder how we know a moon rock is a moon rock and not a worthless stone from your driveway? Well, thanks to solar wind bombardment of the moon's airless surface, its rocks all have more of the rarer O_{17} and O_{18} than earthly rocks do. The specific ratio of rare O_{17} to ordinary O_{16} is distinctive for the moon and for each planet. It's a fingerprint. That's how we know where a particular meteorite comes from.

Most elements are stable; they exist forever. But radioactive elements decay over time, and so do many isotopes of otherwise stable elements. Having extra neutrons makes the nucleus a little weird, so that a proton can eventually fall off and thus change the atom into a different element. You never know when exactly this will happen, only when it's statistically likely. For example, half of the most common uranium turns to lead in 4.5 billion years. This decay period is its *half-life.* Don't wait up for it to happen.

Now back to carbon 14's story. If you think those coronal mass ejections that explosively hurl billions of tons of solar material at us are powerful (and they are), they're nonalcoholic beer compared with the particles that come from outside the solar system. These, too, are simply fragments of broken atoms, but they move at nearly the speed of light and thus pack a wallop. Though made of the same particles, they have been given a totally different name: *cosmic rays.* Supernovas that go kaplooie in distant empires of our galaxy and even other galaxies emit intense sprays of these particles. While the fastest *solar* particles might reach

10,000 miles per second, cosmic rays come in at 180,000 miles per second. Ninety percent of them are protons, which are pretty hefty, and they unleash amazing power when they crash into anything.

Astronauts are brave to begin with, but at least those who merely orbit Earth stay within our protective magnetosphere. However, twenty-seven human beings—the Apollo astronauts between 1969 and 1972—ventured outside the magnetosphere to head to the moon. When they did, they got clobbered with cosmic rays. Each astronaut saw a flash of light like a meteor whiz across his visual field about once a minute. They knew the flashes weren't real because they couldn't be photographed. All the men were concerned, but being longtime pilots, they also knew never to report that anything was wrong with them, no matter what. But when they confided in each other and realized they were all experiencing the bizarre flashes, they informed Mission Control. Experts quickly realized that cosmic rays were continuously ripping through each man's brain, triggering the spurious visual streaks. It doesn't exactly make space seem like a health spa retreat—and it isn't.

Happily for us ground dwellers, cosmic rays are largely absorbed by our atmosphere. Even before they get near Earth, many are kept at a distance by the Sun's own winds. It's a battle between gusts, each with its own magnetic field. The Sun's winds are denser than the more random cosmic rays, and they create a sort of protective bubble around our solar system called the *heliosphere.* But our heliosphere, like a balloon, expands around sunspot maximum, then contracts at minimum. During this latter period, the Sun's fewer and slower particles make for a feebler shield, and twice as many cosmic rays reach us.

After traveling through space for as long as billions of years, cosmic rays reach our atmosphere, where they strike and blow apart air atoms. Stuff flies out, including neutrons, which then slam into other atoms, as in a game of billiards. A likely atom to get hit is nitrogen, since it makes up four-fifths of the atmosphere. The neutron sticks to the nitrogen nucleus but knocks off a proton, and voilà: an atom with six protons instead of seven, which means it has changed into carbon. But it has two more neutrons than in a normal (C_{12}) carbon atom. There we have carbon 14. About seventy tons of it lurks in the air at all times.

We've now established that lower solar activity equals more cosmic rays, which create more C_{14} in our air. Carbon 14 looks, acts, and smells like everyday carbon 12. It quickly combines with oxygen to form carbon dioxide, which is taken in by plants. As a result, all plants have one C_{14} atom for every trillion or so C_{12} atoms. At this point in the tale, you might actually put this book down and say out loud, "So what? Jeez, how long is this moron going to go on about this?" Here's why it's important.

Carbon 14 is unstable. Half of any sample decays in 5,730 years. This is nothing short of wonderful, because it lets us accurately find the age of anything that was ever alive, like the cotton clothing worn by mummies after they had their brains sucked out and were sent packing to the afterlife.

Age is determined like this: Plants take in carbon dioxide, which, to review, means their crunchy green bodies always contain a little C_{14} along with the normal C_{12}; the ratio is a trillion to one. Animals and humans eat the plants, and so we, too, have the same ratio of C_{14} in our bodies. When a plant or animal dies, it stops breathing and no longer takes in new C_{14}, so whatever is present starts to slowly vanish. The regular carbon remains pristine and whole forever, but half the C_{14} is gone in 5,730 years, three-quarters of it is gone in another 5,730 years, and seven-eighths of it is gone in 17,190 years.

So you send a sample of anything you wish to date (we're not talking about girlfriends) to a lab along with their fee of around $600. They analyze it and let you know the item's C_{12}-to-C_{14} ratio, which tells you that, say, 95 percent of your sample's original C_{14} has vanished. You look that up on a chart to determine the item's age. If it died less than six thousand years ago, the lab can determine its age to within forty years. Is that cool or what?

There's more. The amount of C_{14} that everything starts with varies a bit, depending on how active the Sun was during its lifetime. Remember, in periods when the Sun is quiet, more cosmic rays come in, and there's more C_{14} created in our atmosphere. So we can use items whose ages are already known—the cotton burial clothes in archaeologically well-dated Egyptian tombs, for example—and see how much C_{14} was in the air way back then. And thus we can learn how bright the Sun was throughout different periods in history. This works going back about at least forty thousand years.

We can be sure this works by many methods. Tree rings, for example. Some trees are many centuries old, and there's C_{14} in those rings, which we can merely count in order to tell a tree's age. Sure enough, everything cross-checks nicely.

As for finding earthly temperatures throughout history, that's not hard either. One method is drilling ice cores and analyzing the trapped air bubbles. The ratios of certain abundant elements, such as beryllium 10 and deuterium, are sensitive to temperature. We can confirm the accuracy of this method by analyzing cores from just the past century, for which solar irradiance and earthly temperatures are well-known. This and other independent methods, such as sea-level heights and glacier extents, let us determine earthly temperatures going back hundreds of millions of years. So it's not theory. It's not guesswork. Creationists and climate change deniers have it all wrong when they suggest that researchers are just groping around in the dark or coming up with contradictory data.

We also have a fleet of satellites monitoring Earth and the Sun, with instruments that are so clever and sophisticated, I can only shake my head and wonder about the people who created them.

So although some uncertainties exist here and there, this epic saga of determining past earthly temperatures, past solar insolation (sunlight intensity), atmospheric reactions, and the way biological organisms have dealt with these events is now a mostly solved detective story.

CHAPTER 10

Tales of the Invisible

A sunbeam in a winter's day,
Is all the proud and mighty have
Between the cradle and the grave.

—John Dyer, "Grongar Hill," 1726

IT'S SOMETHING YOU might see in a sci-fi movie—the mad scientist theme. The researcher bursts from his lab and wildly shouts, "You've got to believe me! Invisible rays are going through our bodies! They're coming from the Sun! Let go of me!"

Something like that really happened two centuries ago. And nobody would have believed the scientist had he not recently gained global fame by making the most startling discovery of all time.

The background to this story is simple. Like nearly all stars, the Sun emits most of its energy in visible light. Translation: what we see is what we get. Still, there's a goodly amount of solar spillover energy that takes the form of invisible waves. This is a case where "out of sight" is definitely not "out of mind." The unseen realms of sunlight affect our everyday lives big-time.

The best-known spillover energy is detectable by our skin but not our eyes. It's called heat, or *infrared.* Infrared is just like visible light, except its waves are a bit more spread apart, as much as 1/2500 of an inch from crest to crest, or about half the thickness of a human hair.

The discovery of invisible infrared light is justly credited to the great William Herschel, the most famous astronomer of his time (mid-eighteenth through early nineteenth centuries). Here was a true Renaissance man, an accomplished symphonic cellist, oboist, and organist and the composer of twenty-four symphonies (even if all were renowned for inducing sleep). Herschel was also a lens and telescope maker and an astronomer. Ultimately, he built the largest instruments of his day, including a colossal telescope with a mirror four feet wide.

But it was one night when he was using a mere six-inch telescope—the kind many twelve-year-olds receive today as Christmas presents—that he made the most astonishing discovery in human history. At least it was the most unexpected finding of the ages, and it generated incredulity around the world. Herschel was the first person to discover a planet.

To human awareness, there had always been five planets plus Earth, which through most of history was deemed a world apart. This was a fundamental truth, a Bible-level certainty. It had been unchanged since prehistoric times and, more important, had been unquestioned by the ancient Greeks, whose wisdom held sway into the Renaissance. Nobody had ever predicted or even suspected that there might be additional planets—invisible ones.

Herschel's discovery in 1781 not only took the world by surprise, but it produced a kind of insecurity that would be hard to imagine today—sort of like hearing that the Sun is hollow and populated by a race of monkey-people.

The finding propelled Herschel into worldwide celebrity. When he soon discovered two moons orbiting Saturn and two more orbiting Uranus, then a couple of decades later coined the word "asteroid" when those bodies were first found, hardly anyone raised an eyebrow. He was the late eighteenth century's supergeek.

Nineteen years after his discovery of Uranus, Herschel was fooling around with various filters for solar viewing when he noticed that red light seemed to be accompanied by more heat than any other color. Intrigued, he placed a thermometer on a table where cut glass had cast the spectrum of the Sun's colors. In one of those eureka moments most of us get only when we suddenly remember we can mute the commercial, he thought to see if the red section was hotter than the green or blue. It was the simplest imaginable notion, but one nobody had yet entertained. What he found made his jaw drop. The mercury rose the most when the thermometer was held not *in* the light, but *next to* it. The temperature went up when the instrument's bulb was adjacent to the red, on the dark table just before the spectrum began. "Invisible light" must be refracted onto that spot.

In other words, heat is a form of light. And its position among the colors is just before the red begins. Herschel called these *calorific rays. Calor* means "heat," and it took nearly a century before an equally logical term replaced Herschel's: "infrared" (*infra* means "below").

Today we know that infrared waves boast fascinating properties. Visible light comes through a window and warms up items in the room (meaning it makes atoms move a bit faster). This in turn gives off infrared emissions, which cannot easily pass back through the window for an interesting reason. Infrared vibrates at the same rate as glass molecules. This creates a resonance, a sort of jiggling, chaotic barrier to the waves that are trying to get through the glass, like annoyed passengers all angling for the exit on a crowded train. As a result, after the visible light comes through the window, the heat cannot get back out. Thus, the temperature in the room rises. Glass creates an infrared trap! Growers have known this for centuries and exploit it each winter to create tasteless tomato look-alikes. Methane, and to a much lesser degree carbon dioxide, also have this "let some waves pass but block others" capability, which is why we call them *greenhouse gases.* It's also why glass-enclosed rooms and parked cars get so hot.

Infrared also does not like to scatter. Its long waves are penetrating. We see this principle, called *Rayleigh scattering,* in everyday life. Distant mountains appear blue because the blue component of sunlight bounces among the air molecules to create an azure-tinted haze. Green bounces much less, and red less still. Therefore, if you look through red-tinted glasses, which block blue and let red pass through, you can observe those far-off mountains illuminated by only the red component of sunlight. Result: they stand out amazingly sharply. The haze is gone because you're no longer seeing any of the scattered-around blue. This is the principle behind those amber "blue-blocking" sunglasses hawked on late-night TV (which you'd certainly buy if you hadn't foolishly muted the commercial).

If red doesn't scatter much, infrared scatters even less. Exploiting this, astronomers can detect infrared light from the center of our galaxy. This is significant because normal, visible light is scattered around and cannot penetrate all the dusty nebulae, so that we cannot see a thing. No optical telescope can observe the Milky Way's core, twenty-five thousand light-years away and hidden behind Beijing-level pollution, but infrared telescopes essentially feel its heat and create structural maps of what's there.

So heat and infrared come from the Sun and stars, meaning that the Sun is hot. You would have never guessed it, right? To put a real number on it, nearly half the incoming energy from the Sun is infrared.

Beyond the other end of the visible spectrum lies ultraviolet. These rays are the bad boys of the invisible, the rats and roaches of the spectrum. They do *not* get good public relations.

Ultraviolet was discovered in 1801, just one year after Herschel found his calorific (infrared) rays. News of Herschel's stone-simple but world-class observation, which required nothing more than a piece of cut glass and a thermometer, intrigued everyone with a brain, including the Polish scientist Johann Ritter. If invisible rays occupied the area of the light spectrum before the red began, why not more unseen rays beyond the violet? Ritter used paper soaked in silver chloride and exposed it to the space alongside the violet light. Bingo: the paper turned black. Something was there.

Unfortunately for him, Ritter was a procrastinator and postponed publishing his findings. When he did publish them, he blabbered on about unsupported speculations. He was thus duly ignored and only belatedly became recognized as UV's discoverer. That should have won him eventual fame and maybe even fortune, except that he increasingly wrote about occult phenomena, a particular turnoff in that post-Renaissance era. It would be like a modern solar researcher finishing his PowerPoint presentation with personal stories about ghosts. Ritter's peers rolled their eyes and ignored his later scientific work. Growing financial problems and a series of illnesses that plagued his family contributed to his demise. He died an embittered, forgotten man in 1810, at the age of thirty-three.

This dour UV genesis saga, the opposite of Herschel's smiley-face infrared fairy tale, describes rays that make up only a small percentage of sunlight. And of that, only about 1 percent of UV actually strikes the ground. That's because our atmosphere, thankfully, blocks most of it.

All UV wavelengths are not equal. Waves that are just slightly closer together, and thus vibrate a tad faster, carry far greater peril. If you could see them, UV waves 300 nanometers apart closely resemble UV of 320 nanometers, yet the former can burn your skin eighty times faster than the latter. Even the weakest variety, UVA, from 320 to 400 nanometers, which the eye can actually perceive as the color violet, has been decisively linked with malignant melanomas. This longer-wavelength UV easily penetrates the atmosphere.

The main solar villain responsible for sunburns and various skin cancers is UVB, from 280 to 320 nanometers. Fortunately, UVB is mostly blunted by the ozone layer. Even more brutal is the super-rare but fiercest UV type, the mythical UVC, with wavelengths shorter than 280 nanometers. We earthlings never have to deal with UVC because it does not penetrate our atmosphere. Or, if you want to be picky about it, a single photon of it makes it to the ground only every 30 million years.

All solar energy with longer wavelengths than UV—radio waves, microwaves, all the visible colors, and heat itself—may cause atoms to bounce around but are too weak to alter the structure of atoms by making electrons change position. This is why cell phones and leaky microwave ovens, the negative effects of which are feared and publicized in the mass media, can heat up our bodies but cannot induce tumors via genetic changes. The first steps to cancer—alterations in atoms, mutations in long molecules such as DNA, and therefore genetic defects—require that electrons be stripped from their atoms (that is, *ionized*). This is what light with short waves, such as ultraviolet, can accomplish. As noted previously, about 1 percent of the light hitting a sunbather is UV, which translates into a million trillion photons per second, most of which are capable of altering DNA.

Hiding under a hat isn't good enough. One-third of UV scatters in the atmosphere, so it arrives at crazy angles from the sky or from reflections. Remember those distant mountains that look blue because of the photons bouncing among the air molecules? Well, count on far more invisible UV doing the same thing at the same time. That's why sitting in the shade may make you feel as if you're protected from UV, but your cherry red cheeks at day's end will prove that you were not. Blocking the Sun directly is inadequate protection against a burn if you're exposed to a big piece of the sky — or surrounded by snow. UV reflects easily.

Here's something wild: Water absorbs UV when it's calm — and even looks darker then, as if communicating its degree of hazard. Ripply water reflects most of the incoming UV, so a lakeside picnic on a breezy day carries a far greater sunburn risk than when the air is calm.

Vegetation absorbs nearly all UV and reflects almost none. Plants love the stuff. Most varieties absorb at least 90 percent, leaving little to bounce off onto your skin. So your surroundings strongly influence your likelihood of burning. Whereas sand and especially snow throw copious UV your way, grass does not. If you're a fair-skinned blond, plan your picnic for a park rather than a beach.

A cotton shirt blocks 90 percent of UV rays, as does window glass. If you'd burn in two hours outdoors, it would take twenty hours behind glass — meaning it won't happen. Nor will your skin manufacture vitamin D behind glass. The reason: UV waves vibrate about 100 trillion times a second, the same natural rate as the electrons in glass. So UV stimulates a window's electrons to riotous resonances that dissipate energy. (This is similar to the glass-infrared story, since neither rays penetrate windows very well. The distinction: infrared makes entire glass molecules vibrate, while UV does the same thing just to electrons. Visible light does neither, so it passes straight through, uncontested.)

The time of day and season are critically important for UV exposure, too. As the Sun gets lower in the sky, its light has to pass through more air, lowering the percentage of UV that penetrates. In the winter, it's hard to tan or burn no matter how long you sunbathe because of the low Sun angle. But since snow reflects 80 to 90 percent of the ultraviolet striking it, taking to the slopes in November through January carries an increased UV risk that at least partially offsets the usual safety of engaging in outdoor activity during that time.

The biggest factor in UV exposure is the amount of cloudiness. An overcast day means no UV. Australians, with their sunny climate, have a much higher rate of skin cancer than Americans, who generally experience more clouds. But cloud thickness is critical. Thick, dark clouds block everything, while high, thin clouds let almost all the bad stuff through.

Travel influences UV exposure as well, since it increases by 4 percent for every thousand feet you ascend. Translation: Denver's UV level is 20 percent greater than New York City's or Boston's.

Ozone, a peculiar molecule composed of three oxygen atoms, is Earth's primary barrier against dangerous UVB. But it's an amazingly thin and fragile shield. If all the atmospheric ozone rained down to form a layer on Earth's surface, it would be barely as thick as a quarter. Certain chemicals that were once commonplace, such as chlorofluorocarbons (CFCs) and the halon used in fire extinguishers, are powerful ozone destroyers—so much so that UV surface levels grew markedly between the 1970s and the mid-1990s, especially at higher latitudes. At the same time, the incidence of all skin cancers increased. Happily, in one of those eradicate-smallpox-type human success stories, effective international treaties have banned these substances. Although the long-lived molecules will wreak damage for years to come, researchers predict that the UV-blocking ozone will return to normal levels by 2050.

Until then, like mad dogs and Englishmen, we can indeed venture out in the midday Sun. Just not for too long.

Of course, there are worse places than Earth. If on a dare, you went streaking across the moon's surface, you'd suffer immediate consequences (even though there are no laws against this). If we ever have lunar colonies, surely some teenager will find it irresistible to moon the moon.

Studies (accidents, actually) suggest that a person can hold his breath for up to fifteen seconds in a full vacuum without any physical harm and even retain consciousness. There's scientific truth behind the famous scene in *2001: A Space Odyssey* where the astronaut Dave does just that, to outwit the deranged pre-Dell computer HAL. Like Dave, it would be necessary for this teenager to close his eyes tightly, since their corneal water would instantly boil away in the zero pressure. (This helps explain why no actions of this sort will appear in the NASA instruction booklet under "Things to Try When You're Bored.")

All that being said, let's assume that the young man chose one of the rare pleasant moments on the moon, a day after local sunrise, when the night's Slurpee brain-freeze chill of –250°F had gone but before the Sun had heated the lunar surface to its customary high of 230°F. What would be the Sun's effect on his unprotected skin? Without the blocking properties of air, ultraviolet would deliver a painful sunburn after only 90 seconds.

But the story is actually even worse because places such as the moon (and Mars) are outside our protective magnetosphere, which blocks fast-moving solar detritus that has nothing to do with light waves. We're talking about particles, bits of broken atoms. An atom as a whole is magnetically and electrically neutral, so it ignores a magnetic field. But its components, once free, consist of negatively charged, lightweight electrons and positively charged, heavy protons. These are strongly influenced, channeled, guided, or blocked by a planet's magnetic field. Once you venture outside Earth's field, all bets are off. Your HMO would be wise to stop covering you. Suddenly, the Sun's steady stream of subatomic particles, the solar wind — which zooms five hundred times faster than a bullet, even when it's feeling sluggish — continually penetrates your body, smashing into your genes and playing a pinball game with your chromosomes. This isn't good for you.

Brain neurons are destroyed by the nonstop high-speed particles emitted by the Sun. One biologist estimates that during a two-year Mars mission, an astronaut would lose between 13 and 40 percent of his brain. Ouch. (Even us smart people can't afford that.) That greatly exceeds the annual 5 percent neuron necrosis suffered by some Alzheimer's patients.

Let's be clear about terminology. We often use the word "radiation" to mean just about anything that penetrates the body but is unrelated to teenage body piercing. In reality, radiation takes two very different forms. First are electromagnetic waves — forms of light, really. Only the rays whose waves are close together, such as UV and especially X-rays and gamma rays, easily penetrate skin. The second type of radiation is particles, like little bullets. The damage they do is related solely to their heaviness and their speed. It's the old baseball-versus-bullet business. In terms of knockdown power, a baseball moving at 130 miles per hour and a bullet moving at 1,875 miles per hour deliver an equally hard impact. If only bullets didn't break the skin (and create a one-third-inch-wide, foot-long tunnel that invariably bleeds), they would be equally injurious.

When it comes to the Sun's emissions, all light travels at the same speed, 186,282.4 miles per second, but the mass of these photons is negligible. Zero, actually. The Sun's outgoing particles are another story. The heaviest are entire helium nuclei, called *alpha particles,* which are like chubby, slow-moving joggers lugging backpacks filled with protons and neutrons. More common solar detritus consists of single protons and electrons, the latter weighing 1,836 times less than a proton. The Sun's own magnetic field whips these to a speed of a few hundred miles per second as it flings them out into space. During the maximum part of the eleven-year solar cycle, many more particles leave the Sun, and, adding insult to injury, they also move much faster, sometimes reaching 1,000 miles per second.

This is the radiation that might truly prove to be the "stopper" for our sci-fi dreams of colonizing other worlds. Both types, solid particles and electromagnetic waves, quickly sterilize planet and moon surfaces. They are bad for us. Unless colonists are extremely motivated to leave Earth because of serious credit card debt or the promise of fame, or we are collectively lured by the unlikely possibility of finding something super-valuable on Mars, it might be hard to recruit volunteers to go somewhere that gives them a CAT scan's worth of radiation each and every hour. They could, of course, permanently hide out in a lead-shielded shelter on the surface, but that's like going to Hawaii and staying in the hotel lobby.

The bottom line: invisible stuff that streams from the Sun is likely to dictate our space-travel destiny and dreams. It alone—rather than our evolving rocket technology—will probably determine whether we colonize other worlds.

One final fascinating aspect of the Sun's invisible emissions is that the combination of Earth's atmosphere, which blocks most rays and particles, and its magnetosphere, which deflects particles alone, still manages to let in enough to make things interesting. Thanks to the ongoing genetic mutations caused by both types of solar radiation, plus the Sun's modulation of the superfast cosmic rays, fauna and flora are constantly evolving.

A 100-percent-effective radiation block would create a planet with extremely slow-changing life forms. Trilobites and hula-hoopers might still be around. Conversely, a planet with little atmosphere and no magnetosphere, like Mars, has a surface too hostile for *any* life. It's a place where, like the promises of some contractor warranties, nothing much ever happens. Our symbiotic relationship with the Sun's invisible rays is Goldilocks perfect. It creates biological changes, but not at a frenzied rate. It's neither too harsh nor too insipid. It's just right.

But when the Sun goes through one of its hissy fits and hurls extra-strong coronal mass ejections our way (as happens on occasion), and this occurs during one of those few-times-per-million-year milestones when our magnetosphere goes soft during a magnetic field reversal, then all life, except that beneath the sea, sees accelerated mutations, and life evolves rapidly for a while. We have detected such periods in the fossil record.

The Sun doesn't just sustain life. It dictates how fast our own body forms and those around us change into new and improved entities. The solar connection runs to our very genes.

CHAPTER 11

The Sun Brings Death

Well we all shine on
Like the moon and the stars and the sun.

—John Lennon, "Instant Karma!," 1970

LONG BEFORE PEOPLE killed themselves by texting while driving, you could pretty much count on exiting this world via infectious disease or by being the target of one of nature's wild whims. If you got trapped in a place with no water and too much Sun, you were pretty much out of options. Well, what was true in the time of the ancient Greeks remains true today. The Sun still kills a million people a year.

Few of them succumb dramatically, crawling on all fours in the desert. Instead, most pay a long-postponed price for having received a childhood sunburn or for having inherited genes that make them susceptible to melanoma. Still, it's those immediate life-and-death scenarios that rivet us, since the would-be victims struggle in exotic locales far removed from our routine world of malls and traffic tickets.

You can probably recall a favorite doomed-in-the-desert sequence, whether in cinema or in prose. From the original 1965 *Flight of the Phoenix* movie to the heartbreaking true stories in Antoine de Saint-Exupéry's *Wind, Sand and Stars,* death by Sun resembles no other.

For me, the most memorable scenes take place in Slavomir Rawicz's astonishing (but, sadly, out-of-print) autobiographical account, *The Long Walk.* In a century that brought a scourge of psychopathic mass murderers, including Adolf Hitler and Pol Pot, not to mention Hirohito and the Rape of Nanking, Joseph Stalin managed to eke out first place in terms of sheer numbers eliminated. Near the end of the Second World War, Polish army lieutenant Rawicz was arrested and imprisoned by the Soviets for the offense of having his family home near the border and, suspiciously enough, being fluent in Russian. After a year in a miserable far-northern Siberian gulag, he and eight others planned and meticulously executed a nighttime escape on a snowy night and ultimately walked south, without equipment, through all of Siberia, across Mongolia, through the entire Gobi Desert, and over the Himalayas. The few survivors in the group ultimately reached safety and sanctuary in India after almost two years and four thousand miles on foot. Of all their varied ordeals, the Gobi Desert took the greatest toll. Rawicz recounts:

> I tried hard to keep count of the days.... My head ached with the heat. Often the blackest pall of despair settled on me and I felt we were six doomed men toiling inevitably to destruction. With each hopeless dawn the thought

> recurred: who will be next? We were six dried-out travesties of men shuffling, shuffling. The sand seemed to get deeper, more and more reluctant to let our ill-used feet go. When a man stumbled he made a show of getting quickly on his legs again. Quite openly now we examined our ankles for the first sign of swelling, for the warning of death.
>
> Over every arid ridge of hot sand I imagined a tiny stream and after each waterless vista there was always another ridge to keep the hope alive.

The biological mechanism by which the Sun commits murder is complex and varied, but it can easily be prevented with adequate water and proper shading, such as wide-brimmed hats. When these are absent, the effects are often sadistic. According to the *Physician's Desk Reference*, "Prolonged exposure to the sun contributes to sunstroke. When body fluids are not adequately replenished, sun exposure can cause rapid dehydration. Even on mild or overcast days, the sun can have dangerous health effects." The body tries to cool itself with perspiration, and without the replacement of salts and fluids, dehydration becomes the problem. This affects the skin's ability to function properly. So much for the exotic.

Sure, people even in our own plush culture succumb to sunstroke, heatstroke, and Sun poisoning. But we all know that the Sun's more common and realistic threat to life and happiness is skin cancer, the most common cancer of all. Skin cancer is scary and, among most people, misunderstood. Fear of contracting it has made many folks heliophobic, but the best armor is, as always, knowledge. Some of what follows may contradict what you think you know.

First and most familiarly, you are at much greater risk for skin cancer if you have very light skin, blond or red hair, and green or blue eyes. The presence of freckles or a family history of skin cancer matters, too. The greatest predictor for the deadliest form of skin cancer, melanoma, is having lots of moles. If you have none of these risk factors, you enjoy far lower odds of contracting the kind of skin cancer that can kill you, and you should almost surely get much more Sun than you've been allowing yourself, since sunlight is the most powerful *anticancer* agent in existence. More about this soon.

There are two very different kinds of skin cancer, categorized as melanoma and nonmelanoma. If a doctor ever tells you that you have contracted either of the two kinds of nonmelanoma skin cancer, don't panic. The chances are a thousand to one that you will be cured and continue blithely on with your life. These, fortunately, are the most common skin cancers by far. They are so prevalent that more than a million Americans will learn they have a nonmelanoma cancer just this year. Because of this, you often hear mass media warnings to stay out of the Sun, based on the fact that 90 percent of these cancers are attributed to UV exposure.

The next important information is that skin cancer incidence increases according to how high up the Sun is, as experienced from your backyard patio. The rate doubles for every 8- to 10-degree decrease in your home's latitude. (Do you even know your home's latitude? Most people don't. You can find it by scrolling the mouse over your neighborhood on Google Earth.) This means that compared to someone living in Minnesota or Maine, a person six hundred miles farther south, say in Virginia or Kentucky, has twice the risk, since the Sun is always higher there and its ultraviolet is significantly more intense. Go another six hundred miles south, to Texas or Florida, and the skin cancer rate doubles again. Yet another six hundred miles south, in Hawaii, a person is two times two times two, or eight, times more likely to get skin cancer than her relatives back in Wisconsin. This location knowledge is potentially very helpful in deciding whether you need to shun the Sun.

There is more good news. Of those million-plus people who contract a nonmelanoma skin cancer each year, less than a thousand die from it. With a mortality rate of 0.001, it simply isn't very lethal. Put it this way: you have a tenfold greater lifetime chance of getting killed in a car accident than of dying of nonmelanoma skin cancer, even after you've been told you have it.

The most common nonmelanoma skin cancer is basal cell carcinoma. Three-fourths of all skin cancers are of this type. If your doctor informs you that you have it, make a joke, slap her on the back, and change the subject. It likely won't affect your life at all.

You are probably in the doctor's office because you noticed a slightly raised (or even flat) skin growth that is waxy or pearly, or light pink or maybe tan. It might be crusty. Maybe it looks like a sore. She will cut it out using something called Mohs surgery, and the odds are one hundred to one it'll never return. If it does, she'll repeat the procedure. Basal cell carcinoma is so rarely fatal because it never metastasizes.

The second most common skin cancer is squamous cell carcinoma, with 250,000 new cases diagnosed each year. Although a bit more serious than basal cell carcinoma, it spreads slowly and only locally, and has a cure rate of 99 percent.

In sum, the skin cancer scare in terms of sheer numbers mostly revolves around very common but usually easily cured conditions. Excluding melanoma, only around two thousand people per year die in the United States from skin cancer. That's a minor threat to life, similar to that of choking on your lunch. Twenty times more people die annually in traffic accidents.

The statistics change when it comes to melanoma. Many fewer

people contract it—twenty times fewer than those annually diagnosed with nonmelanoma skin cancers—but the peril is high. Of the 71,000 who contract it each year, 8,500 will eventually die from it. The cure rate, currently approaching 90 percent, is a big improvement over the grim fifty-fifty odds of just twenty-five years ago, but it's still nothing to mess with. Early treatment is key.

In terms of what kills us, melanoma misses the top ten cancer mortality list, but some people are far more likely to get it than others. One in thirty-nine men and one in fifty-eight women will develop melanoma in their lifetime, and the incidence is rising by about 5 percent a year. It is mostly a white person's disease. A person is twenty-two times more likely to get it if he's white than if he's Asian, Latino, or especially African American. Indeed, 98 percent of melanoma deaths occur among whites. Unfortunately, people of other races tend to put off seeing a doctor, so mortality rates for them are generally high.

Only 65 percent of melanomas are linked to UV exposure. And in contrast to other skin cancers, which probably depend on your lifetime UV intake, melanomas are associated with burns, especially blistering burns. A single such burn in childhood or adolescence doubles a person's chances of getting melanoma, as do five or more sunburns of any severity at any point in her life. Similarly, those who first went to a tanning salon as a teenager have a 75 percent greater chance of eventually contracting melanoma.

Why is the incidence of melanoma increasing by 5 percent a year? Adil Daud, MD, director of melanoma clinical research at the University of California, San Francisco, is not alone in attributing it to "the increased number of people who are moving to the southern part of the United States, where there's more Sun, . . . and also playing more golf, and going sailing more, and hanging out by the beach." In other words, it's an actual increase, not just due to better detection, and it's primarily caused by our changing lifestyles and the population shift away from the cloudier northern belt, with its lower Sun angles.

Bottom line, if you are white, fair-haired, and light-skinned, and especially if you have a lot of moles, you should wear a wide-brimmed hat and use sunscreen whenever the Sun is strong. Never let your skin burn. But you already know that.

For everyone, especially males age fifty and over, it's a good idea to check your skin periodically for any dark lesions or moles, or have a family member do this. The spots that need to be looked at by your doctor will have (1) an irregular border, (2) an uneven color, (3) an asymmetrical rather than a nice round shape, or (4) a circumference greater than that of a pencil eraser. If you observe any of these things, see your doctor.

So far as Sun exposure is concerned, the key is to avoid getting a burn, especially a bad burn. Unfortunately, fear of skin cancer—and also, no doubt, the vain attempt to prevent wrinkles—has made too many of us hide out from and totally block the Sun. By doing so, we have effectively become a society of underground mole-people. And this, it has now become clear, is killing us as surely as the Sun killed those poor souls through the ages who were trapped in the desert.

CHAPTER 12

The Sun Will Save Your Life

Above all Brother Sun
Who brings us the day and lends us the light.

—Saint Francis of Assisi, "The Song of Brother Sun and All His Creatures," 1225

THE FIRST SCENES in one Sun-tragedy unfolded long before there were written records of any kind. Spurred by events we can only guess at, a human exodus began 50,000 to 70,000 years ago, when our ancestors migrated away from the tropics and the equatorial region's strong sunlight. Immediately, people developed vitamin D deficiencies.

Our bodies make vitamin D when our skin is struck by the Sun's ultraviolet rays. Because UV intensity declines dramatically with lower Sun angles, people in temperate regions, and especially those in even higher latitudes, receive as little as 10 percent of the UV experienced by those near the equator. As our ancestors migrating north developed vitamin D deficiencies, the results were swift and brutal. They were removed from the breeding pool by a cruel Darwinian process: the fetus inside a woman with rickets (a disease resulting from low vitamin D) is unable to emerge from her body, and both die in childbirth.

Within just a few thousand years, natural selection had turned some people's skin white, and they were now able to manufacture ample vitamin D even from the reduced Sun intensity of the higher latitudes. (Dark skin color, called *melanin*, is a sunblock, needed because naked bodies near the equator can suffer from too much ultraviolet exposure.) In North America and northern Europe, the climate is sufficiently warm that their skin was almost fully exposed for more than half the year, and their bodies stored vitamin D in the muscle and fat. A new balance had been restored.

But starting a century ago, everything changed. First, the United States and Europe went from a mostly outdoors agrarian society to a mostly indoors manufacturing one. Then people started driving around in vehicles surrounded by windows. Glass prevents any vitamin D production because it blocks the Sun's UV. When air-conditioning became widely available starting in the late 1950s and then got cheaper in the 1970s, people stopped keeping their windows open. Fixed-pane units became increasingly popular. The only sunlight that reached us in our homes and workplaces came through UV-stopping glass.

The last straw was sunblock. It did not even exist until thirty years ago. The initial UV-reducing creams, which cut exposure only in half, were marketed in the 1950s to promote tanning, not totally screen out ultraviolet rays. Then, in the 1980s, a new product came on the market: sunblock. With SPF (sun protection factor) numbers such as 30 and 45, sunblock essentially stops the body's vitamin D production cold. At the same time, people were advised to cover themselves with these lotions throughout the summer months. Even the medical establishment urged hiding from the Sun as a way to counter skin cancer.

The metamorphosis was complete: we had become like the Morlocks in H. G. Wells's book *The Time Machine,* shielded almost totally from sunlight's UV.

ENTER MODERN VITAMIN D researchers such as John Cannell, MD, executive director of the Vitamin D Council, a nonprofit educational corporation that believes that "many humans are needlessly suffering and dying from Vitamin D Deficiency." Cannell is no ordinary medical doctor. He's no ordinary researcher either. He is a proselytizer, the first in the theater to shout "Fire!" when the smoke appears, while there's still time to get out. And these days, he's very, very passionate. He believes that human beings have unwittingly transformed themselves into something uniquely and self-destructively unnatural.

"We are the first society of cave people," he lamented to me in 2010. "In the development process of creating the skin, nature never dreamed that we'd deliberately avoid the Sun so thoroughly."

What Cannell and a growing legion of researchers are decrying are the past three decades of newspaper and TV scare stories that have made the public afraid of the Sun. The consequence, they believe, is that our blood's natural vitamin D levels are just a tiny fraction of what nature intended. And this is producing an avalanche of horrible consequences that include vastly increased rates of cancer.

That vitamin D is super-important is no longer in doubt. It has become the new needed supplement, recommended increasingly by family doctors and the popular media alike. The March 2010 *Reader's Digest* calls vitamins in general "a scam" and urges people to take no daily supplements whatsoever—with the single exception of 1,000 international units (IU) of vitamin D_3, the form most recommended as a supplement.

This sudden interest has been sparked by a spate of studies strongly indicating that vitamin D is the most powerful anticancer agent ever known. Robert Heaney, MD, of Creighton University, a vitamin D researcher, points to thirty-two randomized trials, the majority of which were strongly positive. For example, in a big study of women whose average age was sixty-two, subjects who were given a large daily vitamin D supplement enjoyed a whopping 60 percent reduction in all kinds of cancers after just four years of treatment compared to a control group.

The skeptical might well wonder how, when cancer typically takes decades to develop, such a huge drop can be detected after just a few years. Heaney believes it's because vitamin D prevents tiny predetectable tumors from growing or spreading. "That's the kind of cancer I'd want to have—one that never grows," he told me in June 2010.

The Canadian Cancer Society raised its vitamin D intake recommendations to 1,000 IU daily in 2009. But Cannell, Heaney, and others think that even this is still way too low.

"I went to a conference and asked all the researchers what they themselves take daily and give to their families," Heaney said. "The average was 5,500 IU daily. There is certainly no danger in doing this, since toxicity cannot arise in under 30,000 IU a day."

Why is this vitamin D craze happening now? It sounds suspiciously familiar—like the antioxidant craze of the 1990s, when everyone was gobbling vitamin E to guard against "free radicals." Or the Linus Pauling–led vitamin C frenzy of the 1970s. Recent studies have shown that all those vitamins have no effect on mortality whatsoever. Indeed, a multivitamin a day now seems to be no better for your health than gobbling a daily Hostess Twinkie.

Perhaps our bodies were not designed to get flooded with vitamins. Or maybe the couple of dozen known minerals and vitamins are only the tip of the health iceberg, and what's important are hundreds, or perhaps thousands, of trace substances of which we are not yet even aware.

Yet it is here, in a discussion of the natural environment in which our bodies were fashioned, that vitamin D makes so much sense. After all, our bodies create it naturally out of the Sun's ultraviolet rays.

Spending just ten minutes in strong sunlight—the kind you get from 11:00 AM to 3:00 PM between April and August—will allow your body to make as much vitamin D as you would get from drinking two hundred glasses of milk. This is astonishing. Asks John Cannell rhetorically, "Why does nature do this so quickly? Nature normally doesn't do this kind of thing."

The implied answer, of course, is that we were designed to have a high and steady level of this vitamin in our bodies. Yet as more and more people are tested, researchers are finding serious vitamin D deficiencies in virtually all of the population of the United States, Canada, and northern Europe. The reason? According to Cannell and the other doctors on the Vitamin D Council, we have been hiding from the Sun for decades.

The results may be even worse than we realize. Many researchers now fear that the explosive increase in autism is a result of pregnant mothers having close to no vitamin D in their bodies and then young babies and infants being similarly shielded from the Sun. The Centers for Disease Control (CDC) says that virtually no infants are getting enough vitamin D. The inadequacy figures, even using the CDC's pre-2011 lower recommendations of what they thought the body should have, was that 90 percent of infants are deficient.

According to Cannell, the highest autism rates occur in areas that have the most clouds and rain, and hence the lowest blood levels of vitamin D. A Swedish study has strongly linked sunlight deprivation with autism. Moreover, blacks, whose vitamin D levels are half those found in whites living at the same latitudes, have twice the autism rates. Conversely, autism is virtually unknown in places such as sunny Somalia, where most people still spend most of their time outdoors. Yet another piece of anecdotal evidence is that autism is one of the very few afflictions that occur at higher rates among the wealthier and more educated— exactly the people most likely to be diligent about sunscreen and more inclined to keep their children indoors.

As we saw in assessing links between earthly events and sunspot fluctuations, it's perilous to assign connections too quickly, and autism in particular is a can of worms. Nonetheless, these early threads should set off alarms: it might be wise for pregnant women and mothers of small children to immediately start exposing themselves and their kids to more sunlight.

When Cannell was in medical school in 1973, he was taught that human breast milk contains little or no vitamin D. "This didn't make sense," he said during a phone conversation with me in 2011. "Why would nature ever deprive a nursing infant of this vital substance?" Then it came to him: "When pregnant women start taking 5,000 international units of vitamin D daily, their milk soon contains enough vitamin D for a breast-feeding baby. So there's the key to how much a woman should naturally be getting every day."

In contrast to all this, and to the great annoyance of physicians and researchers on the Vitamin D Council, the FDA continued to advise only 400 IU of D_3 daily as of early 2011. The agency officially regards most vitamin D studies as "incomplete" or "contradictory" and clearly has taken a cautious, go-slow approach.

In November 2010, the National Academy of Sciences' Institute of Medicine issued its first new recommendations about the vitamin since 1997, and many people were disappointed. The institute did boost its recommended daily amounts to 400 IU for infants, 600 IU for most adults, and 800 IU for those over age seventy. It also said there was no harm in taking up to 10,000 IU daily, although it conservatively adopted 4,000 IU as the official recommended upper limit.

According to Cannell, the new recommendations are still "irrelevant dosages." Michael Holick, MD, of Boston University, another vitamin researcher, agreed, saying that he personally takes 3,000 IU daily.

Cannell told me that the National Academy of Sciences report was a "scandal" and that four physicians had disgustedly resigned from the committee that put out the paper. "Commonsense aspects are totally lacking," he said. "For example, they urge infants to get 400 IU daily, but adults just 600 IU. Yet this vitamin is distributed in muscle and fat. The more you weigh, the more you should be getting. It doesn't make sense."

"Listen," he added, "everyone knows that there is an explosion of childhood cases of autism, asthma, and autoimmune disease. It all began when we took our children out of the Sun. Starting twenty-five years ago, a perfect storm of three events has changed how much sunlight children get. First came the scare of childhood sexual predators in the early eighties, then the fear of skin cancer, and finally the Nintendo and video game craze. Nowadays, kids do not play outdoors. Playgrounds are empty. You're a bad mother if you let your child run around. And it's almost a social services offense if your kid gets a sunburn. Never before have children's brains had to develop in the absence of vitamin D."

Since this is not a medical book, I can only pass on the recommendations of those in the forefront of vitamin D research. Their best advice is to go in the Sun regularly without burning. Wear as little clothing as you can. You know how much Sun you can handle without turning red. Unless you have a very light complexion and blond or red hair, you should be able to expose yourself safely to ten to twenty minutes of strong sunlight at a time. Lie out in the Sun in shorts for five to ten minutes on each side. The key to UV intensity is Sun height. If your shadow is shorter than you are, your body will produce a good amount of vitamin D.

After experiencing twenty minutes of unprotected midday Sun from May to July, or a full hour or more during March, early April, and late August through October, you can certainly use sunblock. The experts say to buy the kind whose active ingredient is either zinc or titanium oxide. Most other kinds will be absorbed by the skin, then enter the bloodstream and circulate. "You might as well drink the stuff," Cannell says disdainfully.

During the low-Sun winter months, you need to spend much more time sunbathing and probably take a vitamin D supplement. The experts are currently urging 2,000 to 3,000 IU daily.

Why not skip the Sun altogether and just pop the pills year-round? Some doctors, including those responsible for the 2010 National Academy of Sciences report, suggest doing exactly that. They figure that you can have it all — nice, high vitamin D serum levels plus no UV exposure, with its skin cancer risk. But others believe that's a bad idea. "Some of my colleagues think D_3 supplements are enough," Cannell says. "But that supposes we know everything. I suspect that we do not know everything. Natural sunlight has to be the preferred route whenever possible."

Everyone should use solar power wisely and not go totally bonkers. There's no need to fry. But whatever extra skin cancer risk we might assume certainly seems to me to be a reasonable price to pay, considering the benefits. It now appears that adequate sunlight-mediated vitamin D might prevent as many as 150,000 cancer deaths a year in the United States alone and also reduce infections, bone problems, and perhaps, though more science is needed, even autism and asthma rates. Of course, on the other side of the balance beam, melanoma causes 8,500 US deaths a year. Every activity from bicycle riding to barroom brawling involves some balancing of risks, and the decision of what trade-offs to make is, of course, yours alone.

Tomorrow is a new day. As the Sun rises, its orange beams will cast magical rays in the morning mist. Is the Sun our enemy or our friend? Will it take our life or save it?

CHAPTER 13

I'm an Aquarius; Trust Me

i who have died am alive again today,
and this is the sun's birthday; this is the birth
day of life and of love and wings

—e. e. cummings, 1950

IN 1982, ALL the planets crowded into one quadrant of the solar system, yanking the Sun one and half Sun widths away from its average position. Would this affect us, with devastating consequences? Astronomers saw this major conjunction coming many years in advance, and nearly all of them shrugged. After all, a far more intense planet alignment had occurred many times before, such as in 1128, with no effect whatsoever.

But not everyone was so sanguine. A pair of renegade astronomers made a wheelbarrowful of money in 1974 by writing a bestseller suggesting that the Sun would respond to this planetary tugging and that this would somehow make the Sun erupt, which would in turn cause California to suffer devastating earthquakes. Like every other predicted Armageddon, nothing happened. And though the suddenly wealthy but chagrined astronomers were widely discredited, not much has really changed in the intervening decades. Naïveté still rules.

While the stone-simple basics of this nearby ball of nuclear fire are nothing short of amazing, most people know just enough to feel confused, like when they read that the Sun travels through the zodiac. We've all heard this and, thanks to the popularity of astrology, can even recite the constellations' names (Leo, Virgo, and so on). But what does that circuit really mean? And is there any actual connection between the Sun's annual month in Taurus and the hordes of stubborn people supposedly born at that time of year? Scientists naturally pooh-pooh all astrological claims, but lots of smart people have believed in a link between personality and Sun position at birth for two thousand years.

Well? Are there any unbiased statistical studies? Does the Sun's "sign" mean *anything?* Are Scorpios royal pains in the you-know-what? Are Aquarians like me really leaders you should follow and trust? Answer: Yes. Trust me.

The widespread paucity of astronomical knowledge among our citizenry becomes vividly clear during the Q-and-A period after my lectures.

Asked one elderly gentleman, "How did the Apollo astronauts manage to steer around all those stars on their way to the moon?"

During a radio call-in show just before the 1994 partial solar eclipse, after I'd warned about the perils of staring at the Sun without eye protection, one woman phoned in with this question: "If the eclipse is so dangerous, why are they having it?"

One high school senior asked, "If the Sun is a star, why can't we see it at night?"

The best way to respond to the low level of public space knowledge is not to recite figures and facts; it's to make the Sun *real.* To make people start afresh, almost like a newborn fawn looking up at that brilliant fire for the first time.

True, it's mostly a ball of yawn-producing generic hydrogen gas. Yet in its central 20 percent, hidden far out of sight, lurks a much smaller sphere, and this one is ineffably astonishing. A whopping 320,000 miles beneath the visible solar surface — forty Earth widths down below — temperatures and pressures are high enough to create a Dantean nightmare of nonstop nuclear fusion. This is where H-bombs go off continuously, a place of inconceivable heat and light. This is the solar system's sanctum sanctorum. This is the Sun's core.

Over time, the solar core's power production keeps getting more intense. The Sun was 30 percent dimmer when Earth was young. Our world wasn't a permanent snowball back then only because the greater greenhouse gases in our early atmosphere trapped just enough heat to compensate for the Sun's youthful anemia.

The future looks even more dramatic. In only another 1.1 billion years, the Sun will give off 10 percent more energy than it does today. Ten percent hotter may not sound like much, sort of like moving from Boston to Atlanta. But in reality, that seemingly small boost will boil away our oceans and sterilize the planet. Game over.

That will be then. Right now, in time spans of centuries, the Sun's output is relatively stable, with only its shortest-wavelength emissions, such as extreme UV and X-rays, varying a hundredfold over its eleven-year cycle. Meantime, the total heat and light it delivers to every square meter of Earth's surface in the tropics (where it's highest up) is like having eighteen 75-watt lightbulbs crammed onto each parcel of land the size of a wall poster. Since our planet has 120 trillion of those square meters, any increase in the Sun's energy output quickly adds up.

The power of the Sun's continuous nuclear fusion is equal to 91 billion megatons of TNT per second. That's 91 billion standard one-megaton H-bombs going off in the time it takes to say "Holy moly." The energy is mostly released as a phalanx of deadly gamma rays and X-rays. As these intense short waves leave the core, they spread out and become diluted. They also plow through dense crowds of atoms, which absorb and then reemit them as photons (light particles) that have a bit less energy and are slightly less lethal. As this process continues, the blinding flood keeps spreading outward. Scientists disagree about the time it takes a typical photon to travel from the Sun's core to its surface. Some say a million years. Some say as little as fifteen thousand. What is not in doubt is that all those atomic absorptions and reemissions ultimately change the energy to about a fifty-fifty mixture of visible light and heat (infrared), and this is what finally escapes from the brilliant sharp-edged surface we call the photosphere. Freed at last, each photon needs just 8 minutes 19 seconds to reach Earth.

Our world orbits this ball of nuclear fire every 365¼ days in an elliptical path. The shape of the ellipse slowly changes over time. These days our orbit closely resembles a circle. The oval is not very squashed, so that we hover 91.5 million miles from the Sun around January 3 each year and stand only 3.4 percent farther, at 94 million miles, on or near every Fourth of July. This small difference is amplified because the solar intensity we feel varies with the *square* of the distance, resulting in a reliable 7 percent annual dimming of the Sun between January and July.

If our planet rotated with its poles straight up and down as

Mercury and Jupiter do, our yearly Sun-distance variation would be the whole story regarding annual temperature changes. But Earth tilts as it travels, so the main factor is where your part of the planet currently points. When the Northern Hemisphere tilts sunward, the midday Sun is high up in the sky, its rays direct and intense: summer. Yet, happily, our summers happen just when Earth is farthest from the Sun. This keeps them from being too hot. The same moderating situation occurs each winter, when Earth is closest to the Sun. Winters are not as cold as they might be.

Or will be. A mere eleven thousand years ago, Earth tilted the opposite way. When that happens again in another eleven thousand years, we'll have our long, dark nights and low, anemic Sun just when the Sun is farthest away. The result: brutally cold winters and scorching-hot summers.

Thanks to such orbital shenanigans (and ignoring man-made global warming), we are now continuing to cool after the last warm period peaked six thousand years ago, just before the pyramids were built. We won't cool much, and may even start to warm up a bit. Bottom line: no new ice age for at least fifty thousand years and probably not until the year AD 130,000, which is when the Chinese plan to devalue the yuan. Only then will ice a mile thick advance all the way to Times Square. Property values will plunge.

Of course, we always feel as if we're not moving as we orbit the Sun. Instead, as we view the Sun from different angles, *it* seems to shift position against the background stars, chugging through the zodiac along a laser line that never varies.

Our planet's motion causes the Sun to move daily by two Sun widths, or 1 degree relative to the stars far behind it. This change in the solar position is not at all obvious, because the Sun's brightness prevents us from seeing what's behind it. Its changing position is also disguised by our clever timekeeping system. Our clocks use a twenty-four-hour solar day instead of the real period of Earth's rotation, which is 23 hours 56 minutes 4.1 seconds. Tacking 4 minutes onto the actual spin period of our planet compensates for our motion around the Sun during this time and lets the Sun appear in the same direction as on the previous occasion when our clocks said 7:00 AM. Thus, we observe the Sun to rise at very nearly the same time as the day before. The trick works beautifully. The Sun keeps being highest when our clocks say noon. This concealment of the Sun's shifting location among the background stars gets further disguised by its rapid daily crossing of the sky caused our own quick spin.

Rotations, orbits, shifts—your head is probably spinning. Let's simplify the whole thing. Pretend our planet didn't spin at all. Then the Sun would just seem to hover in one spot. If it weren't so bright, we could see distant stars far behind it in the background. Still with me? Now, as we slowly orbit around the Sun, we view it from different angles, so that the background stars behind the Sun change very slowly. It's like walking completely around a friend while maintaining eye contact. As you circle her, the background stuff behind her head keeps changing. After you've completed one orbit, the background returns to what it was at the outset.

Every New Year's Day, a beautiful star cluster named M22 in the constellation Sagittarius, which resembles an archer the way I resemble Brad Pitt, floats far behind and just to the right of the Sun, hidden by the blue sky. Astronomers say the Sun is "in" Sagittarius that day.

As Earth chugs along at 66,600 miles per hour, the stars behind the Sun slowly change, even if the blue sky keeps us from seeing this happen. The Sun enters the true constellation Capricornus on January 19, Aquarius on February 16, and so on. Its apparent track against the starry background repeats perfectly every year. This annual solar path is called the *ecliptic,* because even ancient peoples noticed that eclipses happen only when the moon joins the Sun on this imaginary line in the heavens, either together with or exactly opposite it. The Sun's yearly journey along the ecliptic takes it through thirteen constellations, including the little-known Ophiuchus, the Serpent Bearer. (Did anyone ever really get paid to carry snakes? A few who work in zoos or circuses probably still do.) Constellations vary in size, so that the Sun spends a long time traveling through some (six weeks annually in Virgo) but pays only a brief visit to others (six *days* in Scorpius). Given the pros and cons of maidens versus stinging arachnids, this inequality in where the Sun loiters is plausible. Ra and Helios were no fools.

Since the Sun's annual path never changes, you can count on it shining through a particular crack or hole in a rock alignment on the same day every year. Ancient civilizations, with lots of free time on their hands, loved to make this happen. For example, the oldest building in the world, erected even before the pyramids, is Newgrange in Ireland, which annually admits a beam of sunlight to illuminate an inner chamber on December 21. It has marked every winter solstice for the past 5,200 years.

Although the Sun marches along the ecliptic, we know that it is actually Earth that is moving; we merely view the Sun from different angles.

But the Sun really does move. First, no planet strictly orbits the Sun; rather, each circles the spot where its own gravity balances with the Sun's, like the fulcrum of a seesaw where an adult tries to balance with a cat. Since the Sun weighs 333,000 times more than Earth, the fulcrum, or balancing point—called the *barycenter*—is not halfway between them, but rather 333,000 times closer to the Sun's center than to Earth's. The Earth–Sun barycenter lies just beneath the solar surface, and the Sun's core and most of its mass clumsily wobble around this spot once a year thanks to our planet's pull.

Jupiter, 318 times more massive than Earth and 5 times farther away, has such a heavy seesaw presence that its balancing point lies entirely outside the Sun's body. This means, amazingly enough, that Jupiter's twelve-year orbit is not around the Sun, but around an invisible point *near* the Sun. Meanwhile, the Sun, too, performs a small orbit around that spot, which it completes every 11.86 years.

So when we ask where exactly in the sky we must find the Sun, it's not always in the exact middle of the solar system. It shifts by its own diameter in one direction or another, depending mostly on Jupiter's position.

The paranoid among us may imagine that the combined tidal pull of planets during seemingly dramatic alignments will influence Earth and hence us as individuals. But forget that: planets lack both the Sun's great mass and the moon's nearness, the two required ingredients for a juicy tidal effect. To use real numbers, during that 1982 alignment, when every single planet hovered within one quadrant of the solar system, our tides, which normally fluctuate a global average of three feet, rose an extra 1/650 of an inch. That's 0.04 millimeters, or less than the thickness of a human hair. And since animal bodies never experience tides at all, the effect on us as individuals was precisely zero, unless you count the bizarre behavior of running out to buy that "California falls into the sea" book.

Beyond the wobbles induced by the planets circling it, the Sun also moves on its own, taking us along for the ride. It zooms along at 144 miles per second as it circles the center of the Milky Way. Going three hundred times faster than a high-velocity rifle bullet, the Sun and orbiting Earth head toward the summer star Deneb. We would reach it in a million years except that Deneb is going in the same direction at the same speed, so that we'll never actually get to it (making the exercise reminiscent of our chronic futile dieting). We circle the Milky Way's core once every 240 million years.

See it for yourself. Face south at 10:00 PM in mid-September. The center of our galaxy then stands in front of you, a creamy glow about a third of the way up the sky. Forget this exercise if you live in Cleveland or Los Angeles or are standing in the parking lot during a football game. You need a dark sky.

In September, the direction of our flight is straight up, toward Deneb overhead. The stars around us in space, which so beautifully fill the heavens on magical late-summer nights, are, like adjacent horses on a carousel, moving along with us as the galaxy spins.

There is larger-scale motion, too. The Sun and Earth are carried along by the Milky Way's motion through space, and yet we can't say which way we are going, because there is no grid for that vaster scale. We could say we are heading toward the Andromeda galaxy at 70 miles per second. Or is it Andromeda that's coming our way? Or should we split the difference and say that we're approaching each other at 35 miles per second? At cosmic scales, absolute motion becomes meaningless, so we might just shrug and say the Sun is taking us nowhere fast.

WITH SUCH AN intimate, life-sustaining solar connection with our planet and biosphere, are we also linked to the Sun as individuals? Does its position at our birth correlate with our later personality? Or have we been reading the horoscopes for nothing all these years?

Since our focus is the rhythms of the Sun, we cannot ignore its periodic visits to each zodiac constellation and its supposed influence. Even such luminaries as Johannes Kepler believed in this astrological linkage for most of the past two thousand years. And surveys consistently show that more than a third of Americans still believe it.

For decades now, college astronomy courses have included at least one period dedicated to explaining and debunking the supposed personal birth influence of the Sun and planets. The rationale for this use of classroom time is a good one: with thousands of newspapers running Sun-sign astrology columns, people should have some venue in which the light of science addresses this ancient belief system about the Sun's connection to our everyday lives.

I HAVE A singular history with astrology. Far from dismissing it out of hand, I was originally intrigued by it, eager to look for validity in its principles or practice. After all, as a resident of Haight-Ashbury in 1966 and 1967 after I finished my undergraduate studies, and with a philosophical mind-set that carried me to the East for years, I welcomed alternatives to conventional thinking. Adding to the allure was the fact that the two times astrologers did my "chart," the results were predictably flattering. When your horoscope reveals that you're creative and brilliant, what are you supposed to do, tell the astrologer he's full of baloney? (Maybe if astrologers told clients that their horoscopes revealed them to be worthless, obnoxious scum, the popularity of this pseudoscience might wane.)

From the beginning, though, my openness to buy into it was dampened by an obvious and unavoidable reality: astrologers do not speak about the real universe. They place the Sun and planets in imaginary "signs" that do not correspond to the actual constellations, give weight to things such as retrograde motion that are mere tricks of perspective, and ignore factors that should be meaningful, such as the greatly changing distance between Earth and these celestial bodies.

Worse, the origin of astrology is clear-cut superstition. The ancients never performed studies or had control groups. That wasn't their thing. Instead, the same cultures that read the entrails of slaughtered cattle regarded the Sun and planets as gods circling Earth. These were astrology's starting points. Moreover, each planet's supposed "influence" was nothing more than a simple visual association. The god named Mars was reddish, so he was assumed to be linked with fire and blood, and thus war. Mercury moved quickly, hence he must "rule" the trait of changeability.

Attempts to put the universe into boxes based on simple visual or mental constructs have always proved futile. The cosmos resists being snared in the web of our Rorschach musings. Of course, this hasn't kept people from trying, which is why history records endless schemes at linking the Sun and planets with musical notes, geometric shapes, numerological schemes, and myriad other designs.

Knowing as we now do that there is not a drop of blood nor the slightest trace of fire on Mars, it would seem no more capable of causing human aggression than a refrigerator. No wonder astronomers scratch their heads when millions still adjust their lives according to its imagined influence. And this is why astronomers are never astrologers. Nowadays, when someone at a party asks, "What's your sign?" I usually say, "Slippery When Wet." Of course, as an Aquarius, a sign known for skepticism, I can't help but be contemptuous of astrology.

A great many people think there's at least some truth to the notion that your personality correlates with the "sign" the Sun occupied when you were born. Who hasn't overheard someone say, "I can't get along with my boss; he's a Scorpio," or "She's a fussy Virgo."

Astrologers often point to the tides as an example of how celestial bodies can influence us. And the oceans do rise and fall to the rhythms of the moon and Sun. If those bodies can affect the mighty seas, why not the water that makes up most of our own bodies? It seems plausible, but in reality tidal forces are well understood and disprove, rather than support, any lunar or solar effects on the birth of individuals.

Solar and lunar tidal effects do not occur because of these bodies' gravitational pull. Rather, they are caused solely by the difference in their gravity on one side of Earth compared to that on the other. This difference can create slight tractive, or sideways, movements in the seas, which add up. Gravity by itself does nothing, which is why tea doesn't try to climb up the inside of the cup when the moon or Sun passes overhead. Instead, Earth is large enough that there is a slight difference in the solar or lunar gravity on the side facing those bodies versus the far side. This difference doesn't *produce* the tidal effect; it *is* the tidal effect. By contrast, when your five- or six-foot body stands under the Sun or moon, the prevailing gravitational field is the same at your head and at your feet. Thus, your bodily fluids have no inclination to migrate from their customary locations.

But none of this constitutes hard proof that the Sun's position at your birth has no influence on your personality or later life. To find what's real and what isn't, science uses a simple and effective statistical process that is neither mysterious nor particularly difficult.

First, we have available tens of millions of personality evaluations administered by psychologists from 1938 to 1970, when such evaluations were all the rage. Back then, many people had to fill out psychological questionnaires at school, at work, for the military, or when applying for certain jobs. With this enormous data set, any correlation between various personality traits (openness, extroversion, stubbornness, neuroticism, and so on) and birth month would have been found long ago. Instead, all the diverse personalities are scattered randomly throughout the calendar.

Indeed, when researchers combing through mental hospital records noticed a small connection between birth month (January and February) and later onset of schizophrenia, it triggered additional studies; in these larger groups, the original seeming effect vanished. This illustrates how readily scientists will investigate any apparent link between birth month (aka Sun sign) and disease susceptibility.

Sun-sign astrology is easy to evaluate. In 1978 and 1979, Roger Culver of Colorado State University and Philip Ianna of the University of Virginia analyzed and reviewed the mountains of statistical evidence in their books *Sun-Sign Sunset* and *The Gemini Syndrome.* The results were unequivocal: there simply is no such thing as Taurus pigheadedness or Virgo fussiness. More sophisticated studies of the claims of "serious" astrology also have been conducted, with the same results.

Second, since many people believe in "compatible" Sun signs (Libra and Gemini, for example) and "incompatible" ones (Scorpio with Aquarius), two major studies, one in Amsterdam and the other in Michigan, examined marriage and divorce. Each looked at eleven thousand marriages and the same number of divorces.

If we accept that people with "compatible" signs would logically get married more than those with "incompatible" signs, we'd expect more marriages between "compatible" couples. Instead, marriage combinations are totally random. And if we accept that a divorce is one indication of incompatibility, we'd expect more divorces between "incompatible" couples. Instead, these, too, are random.

Such examinations of signs and births have, however, revealed a surprising fact: there are more births in September than in other months. This may indicate a previously unsuspected fertility cycle in humans, related to Earth's position in relation to the Sun. Or, more likely, it may simply be a consequence of widespread leisure time nine months earlier—around the Christmas and New Year's holidays.

Astrology aside, it's not hard to find logical reasons why Earth's orbit around the Sun might produce personality differences. Perhaps a baby born in the spring, whose first few formative months are spent in a carriage outdoors in the fresh air, may be physically or psychologically different from one who was born in November and spent her first six months entirely indoors. Then, too, school enrollment cutoff dates for birthdays, typically September to December, which determine whether a child will forever be among the youngest in his class, could have lasting psychological effects.

Considering all the Sun-related possibilities, what's surprising is the *lack* of any statistical link between birth month and later wellness or personality. This suggests that the Sun affects us all far more as a group than it does each individual. Apparently, as children of the Sun, our destinies are *collectively* intertwined with the fortunes of our home planet and our home star.

CHAPTER 14

Rhythms of Color

Foolery, sir, does walk about the orb like the sun, it shines everywhere.

—William Shakespeare, *Twelfth Night*

NO ORGAN IS more intimately linked to the Sun than the eye. Shaped a bit like the nearest star itself, it was designed to detect the Sun's greatest emissions and then guide us through life via the solar photons bouncing off everything around us.

How it works is not at all intuitive. The vast majority of the centuries in which science was prized above superstition were dominated by the ideas of the Greek physician Galen of Pergamum (ca. 130–200). Residing mostly in Rome, Galen at age twenty-eight was appointed physician to the gladiators, which gave him ample grisly opportunity to peer deep inside the human body to try to ascertain how each organ functioned. Unfortunately for him, dissection of humans was frowned upon both then and for many centuries to come. As a result, Galen acquired many wrong ideas that remained doctrine until the Renaissance, when the sixteenth-century Belgian anatomist Andreas Vesalius pioneered the detailed examination of corpses. Accurate insights into the eye's components and their functions, however, didn't really take off until the seventeenth century, when scientists finally realized it is the retina—not the cornea, as Galen and others believed—that detects light.

Johannes Kepler greatly advanced the understanding of vision by applying his insights about light rays and optics. Kepler was the first to say that the lens of the eye focuses images onto the retina. Supporting him a few decades later, the French philosopher and scientist René Descartes removed an eye from an ox, placed it on a window ledge, and, looking at an opening he'd made in the back of the eye, saw an inverted image of the street outside. (Imagine if it he'd thereby caught his mistress in the arms of his best friend! Why hasn't this been made into a movie?) Descartes confirmed that sunlit images arrive inverted on the retina, after being focused there by the eye's elegantly simple lens.

Jump ahead to the early nineteenth century. The world's long blindness about vision really began to lift when the English physicist Thomas Young explained the wave nature of light. He said that colors are simply waves of different lengths and that the perception of every color comes from a mixture of red, blue, and green.

Soon after, scientists such as Max Schultze in Germany used the microscope to explore the retina in detail. He discovered that two different kinds of cells give us our vision and named them rods and cones because of their shapes. Schultze found that the retinal cones perceive color; by contrast, the rods are color-blind but can detect much fainter images. (In 1938, scientists found that a rod cell can respond to a single photon.)

Vision arises in this book for the simplest possible reason: it is intimately connected with the most obvious and basic of all solar rhythms—the heartbeat of day and night, light and dark. Our planet's spin creates an average daily twelve hours of sunshine, eleven of darkness, and one of dawn-and-dusk twilight, and this pattern is woven into our bodies as inseparably as the cloth threads in our currency. Solar- and lunar-induced circadian and other biological rhythms are numerous and dominant, and some are still mysterious. But we focus here on the single most fundamental and surest aspect of the day-night pattern: our vision.

Obviously, humans are sight-oriented mammals. Rover may mostly trust his olfactory sense when he rips open the kitchen trash. When he then spreads the garbage evenly over the floor, it is not because this appeals to his visual sense of aesthetics, although I suppose we cannot be sure of this. Smell dominates. In humans as in dogs, smell depends on detecting small, simple molecules that can attach to the nasal membrane. Large molecules such as tetracycline or DNA have no smell at all because they're too big to stick to the inside of our noses.

That humans rely mainly on vision is interesting, since we're also the only animals to care deeply about the cosmos beyond Earth, which sends us no other sensory information but light. We cannot taste, smell, or hear the rest of the universe. Until 842 pounds of moon rocks were delivered by the FedEx astronauts between 1969 and 1972, at a cost that wound up being $28,500 per pound, we could touch nothing beyond our planet. Today we're again confined to a situation where virtually 100 percent of our celestial knowledge comes via light. (The exception: probes to comets have brought back samples for analysis.) Even radio telescopes, contrary to popular misconception, detect no sounds whatsoever, but merely the longer varieties of light that we call radio waves.

All light is ultimately solar except for the dim glow of the stars. Moonlight is reflected sunlight. The aurora borealis comes from solar particles exciting the sparse oxygen atoms a hundred miles up. Candlelight and other firelight requires a combustible material such as coal, wood, or oil, all of which are forms of stored sunlight from bygone plants and animals that would never have existed without the Sun.

Today we also create light using electricity, but that current itself comes from burning oil or coal or from hydropower. A world without sunlight would not even have hydro generators: how could dammed lakes get refilled unless sunlight warmed the ground, evaporated water into clouds, and created air motion to transport the droplets back over the higher terrain?

Nuclear power alone is independent of the Sun. Our planet would still have its uranium even if the Sun didn't shine. Ah, but how to extract it from its ore, pitchblende, and then refine it to the required 4 percent concentration? This is an energy-intensive operation; where would *that* electricity come from? It's a Catch-22. You can't create nuclear power without first having a Sun-created energy source to start refining the fuel. Bottom line: all usable light ultimately depends on the Sun. How poetically evolutionary, then, that our eyes see only the colors that the Sun emits most strongly.

As many of us learned in fifth-grade science class, the Sun's white light is merely our retinal/brain response to receiving all the Sun's spectral emissions at the same time. Or, to be precise, seeing white means we're receiving red, blue, and green light simultaneously. Those are light's primary colors. They are totally different from the yellow, cyan, and magenta primary colors of paint.

THE EXPERIENCE OF vision is a symbiotic event. By itself, light has no color or brightness. It is merely a series of pulsing waves of magnetism and simultaneous waves of electricity. The external world that you might imagine is independently "out there" even when no one is watching, is as utterly invisible as radio waves, a complex jumble of various blank energy frequencies. When stimulated by these invisible waves, our retinal cells create a response to a narrow set of predetermined vibrations and send electrical signals at 250 miles per hour up the optic nerve until several hundred billion neurons in the rear of the brain fire in a continuous, complex way. The result is an image perceived in the brain as an experience of "blue" (or whatever). So the "external world" is an internal experience. On their own, colors do not exist. When no one is looking, a sunset has neither color nor an appearance of any sort. It is an invisible mélange of electrical and magnetic pulses.

Nature could have easily designed humans to have a different subjective experience when confronted with the Sun's photons. And some of us do. *Deuteranopes* are the 10 percent of males who lack the green retinal receptor and thus see far fewer colors than the rest of us. To them, shades of red and green can look identical, although they definitely "get" the blues. Deuteranopes can easily run traffic lights if the bulbs are in unfamiliar positions. It turns out that dogs and elephants are deuteranopes, too. It's one of the reasons we should never let them drive.

In sunlight or bright artificial light, we use our best retinal mechanism, those six million cone-shaped cells that deliver full-color vision three times sharper than our prized 1080p high-def TV. Our keenest sight is straight ahead. It's also in the green part of the spectrum, right in the middle of our visual range. Since this is our best color, as well as the one the Sun emits most strongly, it deserves a few moments of our time. Our eyes can distinguish between wavelengths that differ by just one nanometer, but only in the green section of the color palette: human vision can simultaneously detect about fifty different shades of green.

The test for color sensitivity is usually performed on a split screen with slightly different wavelengths displayed on each half.

With less than one nanometer difference, the observer reports a unity of appearance. As slightly different tints are offered on the halves, a critical point arrives where a separate color is sensed: the screen suddenly seems sharply divided in two.

This same sort of setup lets us test animals. Dogs are given a reward if they push their noses onto the part of the screen where the color is different. Such studies show that cats see colors but really don't care about passing the test, so it can take ten thousand tries before the experimenter gets a usable result. Whether in man, monkey, or Maltese, color perception is called *photopic vision.* It functions whenever there is ample light.

This bright-light architecture has plenty of quirks. We see yellow only when we receive an equal dose of green and red light. You'd think confronting a blast of red and green light would give us the sensation of "greenish red," but no such color exists. Yellow *is* our "greenish red," even if it bears no resemblance to either hue.

This contrasts with our subjective experiences of sunlight's other primary colors. We do indeed perceive reddish blue as purple and greenish blue as aquamarine or cyan. Yet thanks to the eye/brain architecture, we cannot gain the subjective experience of a reddish green or a yellowish blue, nor even imagine it.

Another oddity involves sensitivity. The daytime sky is actually violet, but our retinas are so insensitive to this hue at the fringe of the spectrum that we instead see the next most prevalent sky color, blue. Some animals and insects — especially those that can see UV — observe the true violet color of the sky. Birds in particular see far more colors than we do. They can even see ultraviolet, which is why hawks can perceive the glow of mouse urine in a field far below.

Even with all of our bright-light quirks, everything really changes when the Sun sets. When the photon count declines, our pupils expand to triple their previous size, to as much as seven or eight millimeters when we're young. At the same time, photochemical changes in the retina greatly boost its sensitivity, as if our daylight characteristics were now swapped for a different film speed. (Film is a rolled-up strip of chemically treated cellulose that was once used in photography, long ago.) An entirely new process starts working. Normal photopic vision is replaced with our faint-light *scotopic vision.* Before it has fully kicked in — that is, in dim but not truly dark conditions, such as deep twilight or when only a night-light is on — photopic vision still operates, but not well. The colors red and violet at the ends of the spectrum are now seen as gray. Only green objects keep their color. We experience this in moonlight as well. As the light fades further, even the green-blue vanishes. Our eyes are now governed solely by their 120 million rod-shaped cells.

Of all life's solar-induced rhythms, this daily plunge into darkness is the most familiar. And yet such an intimate everyday occurrence causes us to ignore it, the way you no longer hear why your teenager needs more money, but just reach for your wallet while the "reason" unfolds as incomprehensibly as the chirping of crickets. Things can be *too* familiar, so when night vision's characteristics are pointed out, the listener is fascinated, as if hearing about something alien.

It's generally unwise to admit being oblivious to the obvious. "I never noticed you had green eyes!" would be an unproductive comment to a spouse. Yet how many of the following facts regarding the vision you have used every night of your life are familiar to you? I'll list five characteristics of night vision. None of them is technical. Award yourself 20 percent for each one you already know and see if you reach the generous GED passing score of 60 percent.

1. Dim-light vision is dead-drunk slow to get going. Rod cells are lazy; they need repeated stimulation to operate at all. At night

when you turn off your bedroom light, at first you see nothing but total darkness. Within a minute, details of the room begin to emerge. After five minutes, the general features are apparent. Within twenty minutes, if you're still awake, you can see everything you're ever going to see. But if someone clicks on the light for a moment, you're almost back to where you started.

2. Dim-light vision is color-blind. That red sweatshirt and those blue socks you threw over the chair are gray now. The whole room is monochrome. Incidentally, the only animals known to be totally color-blind are owls—hinted at by the fact that their feathers are dull and not vividly hued. Why should they have beautiful colors if none of their friends or relatives can admire them? Their dim-light monochrome acuity is almost infinitely sharper than ours, however.

3. Night vision is very blurry. Normal vision is said to be 20/20 in bright light, although many young people could read the 20/10 line on the Snellen chart. In dim light, however, our best acuity is 20/200. That's legally blind. (If you're strolling along a dark street or park with someone on a first date, you could play the "sympathy angle" by telling her you're legally blind. At that moment, you wouldn't be lying.) Test it tonight. When your bedroom light is on, every fiber of the carpet or grain in the wood floor stands out. When the light is off and conditions are very dim (the only light is coming from the crack under the door or a streetlight veiled by the window curtain), the rug or floor looks uniform, and all detail is gone. The same is true of blades of grass when you're outdoors at night away from artificial streetlamps. There's no sharpness at all.

4a. In dim light, you have a blind spot in the very middle of your vision. It's straight ahead. This is not like the benign off-center blind spot of bright photopic vision, caused by the fovea where the optic nerve meets the retina. That one's benign, because the blind spot of one eye never matches the blind spot of the other. By contrast, the dim-vision blind spot is central, it's the same in both eyes, and it's large — a full degree across. That's twice the size of the moon. This happens because only cone cells are located in the center of the eye, which is why we see best in bright light by looking directly at the object of interest.

4b. At night, we perceive the best faint detail when looking slightly off to the side, while keeping our attention on the subject. Astronomers have known this for centuries. An observer with normal vision can resolve the many individual stars in the Beehive Cluster in the constellation Cancer, but only by using such averted vision. When stared at directly, the cluster is a blurry blob.

5. Deep reds do not show up at all in dim light. It's not merely that they become gray; it's worse than that — they vanish altogether. Rod cells simply cannot perceive wavelengths longer than 630 nanometers, which unfortunately includes the most common color in the universe, the red glow of excited hydrogen that is the calling card of nebulae. One fascinating demonstration involves Christmas lights on a rheostat. As the brightness is turned way down, the blue, yellow, orange, and green bulbs suddenly reach a dim point where they go gray. But the deep red lights never turn gray. When they get sufficiently faint, they simply vanish.

We experience these realities of vision every waking hour of our lives. (Well, maybe not that holiday-lights-on-a-rheostat business.) They are too common to get noticed. It's time, then, to turn to a phenomenon discovered half a century ago — one that is, in a sense, exactly the opposite of vision. This is a solar entity utterly unfelt and unseen, yet one that is absolutely everywhere.

CHAPTER 15

Particle Man

Love is the shadow that ripens the wine
Set the controls for the heart of the sun
The heart of the sun
The heart of the sun.

—Roger Waters, "Set the Controls for the Heart of the Sun"

THE 1930S WERE a bountiful time to be an astronomer or, indeed, any kind of physicist. Juicy discoveries were being plucked yearly like ripe apples. Holding a gunnysack open under this prolific tree was Wolfgang Pauli, one of the most brilliant of a coterie of oddball European theorists.

Born in Vienna, Pauli quickly gained fame as a cutting-edge quantum theorist in Austria and Switzerland before eventually immigrating to the United States. His keen mind was widely admired by scientists, and yet his was no life to be envied. His personal difficulties began at age twenty-seven, when his mother, with whom he was very close, committed suicide. The next year, 1928, his father married a woman Pauli routinely called "the evil stepmother." And although his career in Zurich was soaring, his personal life was going in the opposite direction. He got married in 1929 but was divorced in less than a year. A womanizer and depressive neurotic, Pauli grew increasingly distraught, turned to drink, and then sought help as a longtime patient at the clinic of psychoanalyst Carl Jung. He ultimately developed a friendship with Jung himself. The two shared a love of numerology, among other things, and Jung eventually wrote a major book based on hundreds of Pauli's dreams.

In fact, much of Pauli's life was dreamlike. For example, whenever he stepped into a lab, it seemed as though equipment failed or broke, as if some demonic energy was at work. Among physicists, this became known as the *Pauli effect.*

Pauli obsessed over newfound revelations in physics whenever the value of some physical constant seemed to have no rhyme or reason. Of particular frustration was *alpha,* the fine-structure constant, which shows the strength of the interactions of the electromagnetic force associated with light, magnetism, electricity, motors, and everything else in modern technology. Alpha's value is 1/137, and this somehow bothered everybody, despite the fact that other constants, such as the force of gravity and the speed of light, have values that are just as capricious. Alpha's appeal arose partly because it's a "dimensionless constant," which means that it doesn't change, no matter what units are used. Switch from metric to imperial units, and its value is still 1/137. Pauli had disquieting dreams about this constant. In later years, when someone asked him what, if allowed a single question, he would ask God, he said without hesitation, "Why 1/137?"

But it was by observing atomic nuclei in 1930 that Pauli made a discovery that yielded the keen prediction that interests us here. It was only one of many that helped propel him toward the Nobel Prize in Physics, which he received in 1945.

At the center of all atoms except hydrogen lurk neutrons, which sit inertly like Jabba the Hut. They are nature's heaviest stable particles. They live forever. But this happy longevity strangely vanishes if a neutron leaves its atom; then it's a goner. A lone neutron decays in about eleven minutes.

A loose neutron is a loose cannon that goes shazam and vanishes in a puff of smoke, turning into a proton and an electron. The slightly odd way this detritus flies off, like a defective firework, made Pauli realize that another object must be present in the scattering debris. Whatever it was, its influence must be tiny. Nor did it yank at those charged particles in a manner suggesting that it had any sort of electrical charge of its own. At first Pauli named this unseen object "neutron," but this name was soon assigned to the newly found neutron itself. The American physicist Enrico Fermi suggested the name *neutrino,* or "little neutral one," for Pauli's predicted particle, and the name stuck.

For decades, no one had any idea how to detect Pauli's neutrino, but that didn't stop it from gaining fame. The evolving theories explaining the fusion process that powers the Sun needed the neutrino to exist in order to make sense of the total energy and mass that were so obviously being created. If it didn't exist, solar physicists would have to go back to the drawing board. (It turns out that a whopping 3 percent of the Sun's energy is in the form of neutrinos.)

A full quarter century passed after Wolfgang Pauli's prediction before someone finally figured out how to find the mythical neutrino. It was a scientific triumph, even if it never was discussed around office water coolers. The problem had not been one of scarcity. The Sun's fusion process releases countless neutrinos, which are one of nature's very few particles that cannot be broken down into anything smaller, as far as we know. The universe's only other fundamental particles are quarks and electrons. (Quarks, which come in threes, stick together to create the neutrons and protons in every atom.)

Neutrinos are everywhere, like roaches in Rio. Indeed, it turns out that neutrinos are the most abundant items in the universe. Each second, twenty trillion neutrinos fly through our brains. But we'll never think strange thoughts because of them, and they almost never meddle with our organs.

Neutrinos from the Sun are far and away the most prevalent things in our lives; nothing else even comes close. And yet they remained undetected until 1968, the year the Beatles went to India (although the two events are generally regarded as unrelated). Although the neutrino was predicted to explain odd atomic behavior involving the neutron—whose name is so similar that the two are often confused—the zippy, virtually massless neutrino bears no resemblance to the obese, stay-at-home-with-a-beer neutron.

IN THE SUN'S FUSION FURNACE, neutrinos whiz away at essentially the speed of light. Instead of colliding or in any way interacting with atoms, photons, or other particles, they just pass through everything they encounter. They act as if the universe is invisible to them. Once in a blue moon, a neutrino will change an atom into something else, but it's rare.

By day, solar neutrinos hit your hair, pass through your head, and keep on going. They zip right through your body and continue into the ground, through the entire planet, and out the other side without slowing down in the slightest. At night, an equal number of neutrinos arrive from the Sun in Earth's opposite hemisphere, whoosh through the planet in one-twentieth of a second, enter your sleeping body from below, and exit upward, continuing on into space.

This sounds disconcerting, but fortunately, like the one million dust mites in your mattress, neutrinos are harmless. The chances that a neutrino will strike and alter even one of your body's seven octillion atoms anytime in your entire life are one in twelve thousand. Count on it happening to the other guy. You'd need a wall of lead a light-year thick to stop the average neutrino. And where would you get one of those, except maybe on Craigslist?

It was physicist Ray Davis who finally figured out how to detect and count these ghostly objects in 1968. He probably got the idea after dripping blue cheese dressing on his favorite jacket, because the experiment called for a huge vat (100,000 gallons) of cleaning fluid. It took ten railcars of the stuff to fill the specially constructed tank. Making it even more challenging, Davis placed the tank a mile underground in the abandoned Homestake Gold Mine in Lead, South Dakota. Only bats and neutrinos could get near his setup. He figured that a million pounds of tetrachloroethylene would have enough chlorine atoms that a neutrino would occasionally change one to a detectable form of argon. The bizarre apparatus—science at its most spanking clean—actually worked, and Davis deserved the 2002 Nobel Prize in Physics he won just four years before his death.

Why, you may wonder, did it take the Nobel Committee thirty-four years to give Davis the prize? Because no one was sure whether the experiment really worked. It did find solar neutrinos, but the Homestake Mine apparatus detected less than half the expected number.

When the results were first announced, no one was happy. Were we wrong about the Sun's fusion process? Had Davis made some mistake? Or was the neutrino itself a different animal from what everyone expected? Why were there so few of them? This "missing neutrino" problem made astronomers twitch and squirm for a quarter century.

Actually, from the beginning, when Davis first planned to get involved with neutrinos in 1946, he knew that the most common solar fusion process, in which hydrogen changes to helium, creates neutrinos with too little energy to be detected by his method. Happily, the Sun also produces a bit of its energy a different way, by fusing helium into carbon (a process that will get more and more important in the Sun's old age). It's these higher-energy neutrinos that the cleaning fluid business could detect. But only about one-third of the expected number of neutrinos came in. Why? The Nobel folks were definitely not going to award Davis anything until they knew the answer. Something was seriously screwy here.

The world's physicists caught the solar neutrino fever big-time. They built a detector in Russia using the entire world's inventory of gallium. They built another in Japan that used a reservoir's worth of ultrapure, continually cleansed water. In 1999, they built a clever device in an Ontario nickel mine using a thousand tons of "heavy water" (its hydrogen atoms are isotopes having one neutron instead of the usual none). A "bottle" of the stuff—a massive rocket ship–like container holding two million pounds of the odd liquid—was suspended in a lake of more ordinary pure water, and it was all surrounded by 9,500 light-detecting photomultiplier tubes, wired together like the ultimate home theater. They were looking for flashes of Cerenkov radiation, the eerie blue glow created when anything travels faster than light. Incoming neutrinos, they figured, would produce this effect.

Clearly, the missing neutrino quest was no small potatoes. Solving this Sun problem became an international crusade reminiscent of the eighteenth-century astronomical unit obsession, except this time the costs were even higher, and no one cared about the budget.

The solution was finally found in 2001 in the Ontario nickel mine. It was nothing short of bizarre.

Neutrinos come in three different varieties, which we'd always suspected. What we didn't know was that they change from one kind to another as they fly through space. Imagine that you wake up one morning to find you are a rhesus monkey. After a few minutes, your simian self vanishes, and suddenly you're a schnauzer. A minute later, just as you are about to lap up your water and kibble, your new canine persona vanishes, and you're back to being you. The relief is short-lived, since, poof, in another minute you're that monkey again. And the whole process keeps repeating forever. Well, this is what neutrinos do, and each variety is very different from the other two.

Since the original Homestake Gold Mine experiment could detect only one type of neutrino, Davis kept getting a "short count." Now, with the Ontario results, his 1968 data made sense. He hadn't screwed up after all. Just one year after the Ontario findings were published, the Nobel Committee gave Davis the prize, and the world's solar astronomers could breathe again. The missing neutrino problem had been solved. It turns out we'd had a perfect handle on how the Sun shines all along.

As a side note, knowing now that the universe is packed with neutrinos that seldom interact with anything, the idea of dark matter no longer seems so far-fetched. Those gravity-producing, unseen objects make our Milky Way spin oddly. They are presumably more massive than neutrinos but meddle even less with our bodies, our selves, and our planet than do neutrinos. As before, the trick is to find a way to detect something that's not only invisible but also has little or no effect on the things around it, or on us. (Has science gotten just a wee bit more challenging than when we played with magnets in second grade, or what?)

Solar neutrinos cannot harm us, but the same cannot be said of the other solar particles that zip through our bodies. Actually, some of *these* tiny bullets come from much farther away—from distant empires of the galaxy. We each get struck by a heavy neutron or a proton about once a second. The ones from the Sun are far more numerous but move more slowly and hence are less penetrating. The ones called cosmic rays travel at a healthy fraction of the speed of light and pack a punch. A good charge-coupled device (CCD) camcorder in total darkness detects some of these as random flashes of light. The frequency at which they strike us depends on our elevation and whether we're close to the equator, where Earth's magnetosphere does a better job of blocking them. Compared with the cosmic ray intensity in New York City, you'd receive 10 percent less of this radiation exposure if you lived in Miami. By contrast, living high up in Colorado means that far more particles from the Sun and beyond would penetrate your favorite organs. Compared with Brooklyn, your exposure would be 388 percent greater in Denver. Hope all those pretty white Coors mountains are worth it.

One of the most fascinating of these body piercers is the *muon*. The muon is not a weightless neutrino. It is two hundred times heavier than an electron, and it packs a wallop. Muons damage DNA and can cause cancer. They are responsible for some of the spontaneous tumors that have plagued the human race since long before we started charring wieners. About one hundred muons tear through your body every second. Sure, that's a trillion times fewer than in the simultaneous neutrino invasion, but muons can be infinitely more troublesome.

Muons are not sent to your body from the Sun. Nor do they come from anywhere on Earth. Actually, they do not come from space either. Rather, they're created, like the monster Dr. Frankenstein's HMO refused to pay to sedate.

Muons are produced when high-energy cosmic rays—90 percent of them protons—arrive from distant parts of the galaxy or sometimes the Sun after one of its most powerful kinds of eruptions, a coronal mass ejection. These high-speed bullets break apart atoms thirty-five miles up, resulting in a chain reaction of subatomic debris that creates muons, which continue traveling downward at the same near–light speed as the original cosmic particle. The solar wind and solar magnetic field either deflects them away from Earth or allows more to arrive. From 2006 to 2009, the Sun let more cosmic rays reach Earth than at any time in the past century. From 2011 to 2015, the Sun is expected to get back on guard duty and repel them.

Muons were discovered in 1936 by Carl Anderson during his studies of radiation from space. He observed an unknown particle that acted strangely in a magnetic field, not curving as abruptly as an electron. He correctly figured that it must have a negative charge but also that it must be much more massive than an electron, which kept it from being as agile. He rightly assumed that its mass lies somewhere between the obese proton and the skimpy electron. He also found it to be unstable. We now know that after just two-millionths of a second, a muon goes poof and turns into an electron and two neutrinos, neither of which can harm a fly.

No theory had anticipated the muon's existence. Its discovery was so unexpected and disconcerting that future Nobel laureate Isidor I. Rabi actually exclaimed, "Who ordered *that?*"

The muon's short life span serves as a perfect demonstration of Einstein's relativity theory. See, if it decays and vanishes in just two microseconds, that's only enough time for it to travel seven city blocks, even at nearly the speed of light. Yet it somehow manages to traverse the thirty-five miles down to your living room and rip through your body. What gives?

Relativity means just that: time and distance are relative. They are different to observers in separate reference frames. To a muon moving at almost the speed of light, time passes normally, but distance shrinks dramatically. To a muon, your body is not located thirty-five miles down below, but just a few yards away. It still decays in two microseconds, but that's plenty of time to cover the short distance to your gallbladder.

We do not travel at such high speeds and thus have a different reference frame zip code. To us, Earth's atmosphere is *not* a few yards thick, and that muon starts out at an altitude of thirty-five miles. However, we observe that the muon's time is running very slowly. Instead of decaying harmlessly in just two microseconds, the way it does when it's at rest in a laboratory, its high-speed life ironically unfolds in slow motion. Its decay is delayed so much, it remains intact all the way down.

We see distance remaining unaltered but the muon's time unfolding at a snail's pace. The muon feels a normal passage of time but observes distances as wildly shrunken. Which one of us is right? That was Einstein's whole point: there is no correct time or distance between anything and anything else. Using either reference frame, ours or the muon's, the result is the same: the muon penetrates your home like a cat burglar and does its mischief.

MIND-STRETCHING SIDE CONSEQUENCES accompany both of these Sun-associated particles, neutrinos and muons, as they whiz through our bodies nonstop.

Wolfgang Pauli left the United States after World War II to spend his final decade in Switzerland. He lived long enough (barely) to see his predicted particle, the neutrino, discovered. The same satisfaction was not granted him regarding his other obsessions.

In 1958, as he was lying in a hospital bed, dying of pancreatic cancer, a visiting physicist commented on his room number. It was 137, that bewildering alpha number.

"Yes, I know!" the fifty-eight-year-old Pauli said with a smile.

We will never know whether he actually got to ask God about it.

CHAPTER 16

Totality: The Impossible Coincidence

Sunshine on my shoulders makes me happy...
Sunshine almost always makes me high.

—John Denver, "Sunshine on My Shoulders," 1973

IN THE THIRTEENTH CENTURY, an army of laborers and sculptors converged on a barren part of northeastern India, in what is now the state of Orissa, to build the world's most astonishing monument to the Sun. When completed, the Konarak Sun Temple stood as tall as a thirty-story skyscraper.

Still in good condition, the temple is the chariot of Surya ("Sun" in Sanskrit and Hindi). The enormous spectacle appears to roll on two dozen huge carved wheels, pulled by seven horses, and its entrance is guarded by lions. One of the most amazing works of humankind, this structure was the goal of countless pilgrimages by Sun worshippers from a thousand miles around. Visiting it would have been a dizzying experience. At the temple's entrance, dancers performed among the intricately carved geometric patterns. Every inch of the temple itself is covered with exquisite sculptures, many of them depicting erotic scenes of lovemaking so explicit they could not be shown today on the most liberal cable TV network. But beyond this titillating and unexpected oddity in a very modest nation, every morsel of the temple's art, ranging in size from the enormous to the miniature, unfolds with beauty and grace. The famous Bengali poet Rabindranath Tagore said of Konarak, "Here the language of stone surpasses the language of man."

In the sixteenth century, the country fell to the Mogul Empire, and the invading Muslims set about destroying Hindu temples wholesale. How did the marauders, who controlled India for the next two centuries, miss Konarak? Perhaps its extreme remoteness saved it. By 1779, Hindu priests had spirited away the innermost holy relics for safekeeping. A mere twenty years later, Konarak and its grounds had reverted to dense forest, empty of any inhabitants, blasted by sand, roamed by wild animals, and visited only by pirates.

Except for the pirates, not much had changed by 1971, when a series of fortunate circumstances brought me to Konarak to stare at it in admiration and disbelief. I was amazed, too, by the lack of visitors and any surrounding population centers in this rapidly growing country. Most fascinating of all, my astronomy books told me that the rare narrow path of the totally eclipsed Sun would pass directly over the temple nine years hence, in 1980. It would be the first such eclipse there since 1688. How astonishing that the rare magic of solar totality would come to this spot, of all places, in my lifetime. But would my own unknown future finances allow me to be in sync with this near-mythical event?

As it turned out, the intervening years brought me a marriage to a beautiful Italian woman, who loved astronomy and the exotic. As no magazine I was associated with had the good sense to finance my trip, we merely circled the date on the calendar, watched as it approached, and saved as much as we could.

On February 16, 1980, we found ourselves in a town forty miles from Konarak. The car we were riding in stopped, and we rolled down the windows and listened. Total silence. The town was deserted. This part of Orissa, two hundred miles south of Calcutta, was normally cacophonous. But now no blaring horns or animated fruit sellers disturbed the stillness. Every house was shuttered. The streets looked like a scene from an ultralow-budget sci-fi movie. Restarting the engine, my Indian companions—a young reporter from Delhi and an elderly astrophysicist—said they were not at all surprised. "Villagers think that eclipses of the Sun are extremely dangerous," the astronomer explained. "They bathe before and after, and any food prepared during the event is regarded as poison." Later I learned that pregnant women—who appeared to account for most of the population—had to take particular care to keep their eyes tightly shut, lest the fetus be malformed.

An hour later, we reached the thirteenth-century temple with its X-rated frescoes. But like a pointillist painting, the true nature of this solar monument was clear only from a distance: the colossal horse-drawn chariot sitting like a mirage in the middle of the forlorn desert, still undisturbed and untouched by solar totality since the Mogul Empire. This was indeed a Hollywood-perfect backdrop for the natural spectacle about to unfold—the most amazing apparition the human eye could ever see.

* * *

WHEN I WAS A KID, nothing seemed more mythically wondrous than a total solar eclipse. In *A Connecticut Yankee in King Arthur's Court,* Mark Twain's hero escapes death by bamboozling the natives with a wave of his hand, seemingly making the Sun go dark. To me, the most riveting possibility on that page wasn't that captured adventurer possibly being burned at the stake, but the Sun possibly turning black. *Can that really happen?*

The rarity of a total eclipse adds to its allure. For any given place on Earth, a totality appears just once every 375 years. If it's cloudy, you have to wait another 375 years. But that's just the average. Here and there, a few odd places enjoy two totalities in a single decade, while others must cool their heels for more than a millennium. In the United States, no major metropolis has seen a total solar eclipse since New York City did in 1925, unless you count the clouded-over sunrise event that disappointed Boston in 1959.

During that Roaring Twenties Big Apple totality, the width of the path of darkness ran from Albany, upstate, down through the Bronx and Harlem, and ended unceremoniously at 86th Street, near an eatery that would one day be famous for hot dogs and papaya drinks. People south of that subway stop stood in daylight: no stars out, no mind-numbing glimpse of the solar corona, no hot-pink flares shooting from the Sun's limb. Volunteers were dispatched to each street so that scientists could later know the precise location of the edge of the moon's shadow. The next day, a newspaper reporter, having watched the disappearing Sun's final dazzling pinpoint, described it as a "diamond ring"—a term that has since been fully incorporated into eclipse-speak.

Firsthand accounts from that eclipse and others did nothing to cure my adolescent eclipse fever. All descriptions were unanimous. Animals went nuts. People howled or wept. Flames of nuclear fire visibly erupted like geysers from the Sun's edge. Shimmering dark lines covered the ground. Could it get any weirder?

I grabbed the first chance I could to see the spectacle. On March 7, 1970, the moon's one-hundred-mile-wide shadow was forecast to cross over Virginia Beach before heading out to sea and then ultimately passing over the enchanted island of Nantucket, off Cape Cod. As a naive teenager, I did not yet know that nature guards such wonders by erecting obstacles, usually an overcast but sometimes a more bizarre impediment. Famed NASA eclipse expert Fred Espenak later told me that he once ran panting at full frantic speed along a dirt road in Africa, trying (and succeeding) to keep the eclipsed Sun visible through a tiny opening between moving clouds.

In my case, the rare forecast of pristine blue skies over the entire Northeast made both potential observing sites equally attractive, creating a dilemma. Newly licensed to drive, a couple of friends and I chose the fifteen-hour road trip south to Virginia. Good thing. It turned out that all the Nantucket ferries were sold-out. We would have been reduced to standing at the dock with the disappointed crowds, watching a 99.9 percent eclipse from the mainland. Missing totality by less than 1 percent may sound like a reasonable experience, nothing to complain about, but it's actually no better than *almost* falling in love or *almost* seeing the pyramids. Only full totality produces the astonishing phenomena nature provides on no other occasion on this planet nor on any other in the known universe.

Dawn found us reaching the Chesapeake Bay Bridge-Tunnel, where our charts showed we were now within the ribbon of totality that would arrive a few hours later. I said to the toll taker when we stopped at her booth, "You're so lucky. We've driven all night to get here, and you only have to look up."

"Not me!" she insisted, as if my suggestion was a sign of derangement. "You'd go blind. You won't catch me watching it."

Here was the most amazing visual experience of her lifetime, bar none, and she was going to turn away, as if heeding the lesson learned by Lot's wife. Years later, as an eclipse lecturer, I saw this same behavior over and over, as in that Indian town. Even American schools got into the ostrich act. Perhaps intimidated by attorneys, some teachers kept the shades drawn, ignorant of the fact that only the hour-long *partial* phase requires eye protection. When fully eclipsed, the Sun can be safely observed directly, even through binoculars.

IF TOTALITY IS A KNOCKOUT, the strange science behind it is no less so. Indeed, here is a case where the science spills over into something approaching witchcraft.

Many folks already look to astronomy to help assess whether the cosmos is dumb versus divine. Unhelpfully, the universe does not generally oblige. True, for centuries only a divine plan seemed capable of explaining why the moon spins in the precise same period in which it orbits Earth, so that it always keeps one side facing us. But tidal forces perfectly account for this apparent oddity, and, it turns out, the moons of all the other planets perform the same synchronous rotation. Every moon in the solar system forever presents just one face to its parent planet. No need to invoke the Deity.

Eclipses are different. They truly fit the "amazing coincidence or divine plan" category. How else to explain that the moon is four hundred times smaller than the Sun but also four hundred times nearer to us? This makes the only two disks in our sky appear the same size. That would not be the case if either were larger, smaller, nearer, or farther away. Such an astounding coincidence also lets the Sun's inner corona and prominences stand visible all around the moon's inky limb.

Making it even eerier, the moon wasn't always where it is now. It really just arrived at the "sweet spot." It's been departing from us ever since its creation four billion years ago, after we were whacked by a Mars-size body that sent white-hot debris arcing into the sky. Spiraling away at the rate of one and a half inches a year, the moon will eventually appear too small to cover the Sun. In another seventy million years, an eyeblink on the cosmic scale, eclipses will be over for keeps. The era of these spectacles corresponds with the brief time humans will occupy this planet.

For early cultures that regarded celestial phenomena as magical to begin with, eclipses lay entirely off the weirdness scale. Some, like the Chinese and the Babylonians, were obsessive enough to make astoundingly accurate observations that ultimately gave their priests the power of prediction.

When we talk about various solar rhythms, it's hardly original behavior, since human fascination with recurring celestial events predates the Parthenon and even the Sphinx. Back then, no one suspected that the Sun itself changes; indeed, its stability was its calling card. Only something both powerful and relatively dependable could be "God," since nothing's worse than a twitchy deity. For early civilizations, the concept of divine solar rhythms lay exclusively in how the Sun repeated its positions at regular intervals, plus its role in the rare and bizarre eclipses.

The ancients easily noticed that eclipses come in several different flavors. They can be solar or lunar, partial or total, short or long; they can happen in the warm or cold season, with the Sun high or low. Moreover, the moon's shadow can move along the ground toward the north, northeast, east, southeast, or south. The Babylonians noticed that an eclipse with all the same specific traits will exactly repeat after 18 years plus 11⅓ days. This observation was amazingly perspicacious, especially since that "1/3 day" business ensured that the next eclipse would be best seen (or maybe *only* seen) in an entirely different region of the world. The Babylonians called the period between identical eclipses the *saros.* The ancient Greeks loved that word (and concept) so much that they embraced it without even translating it into their own language.

The saros's "1/3 day" has annoying ongoing implications, for it means that Earth turns through 120 degrees of longitude for each subsequent event. For the same type of eclipse to appear again in an observer's region, she has to wait while eclipses work their way around the world like a set of gears, which requires three saroses. Earth then completes a full rotation, bringing the spectacle back to the original location, albeit with a slight shift. This time period of three saroses, an important observational interval with the Hogwarts-like name of *exeligmos,* amounts to 54 years plus 1 month. As this surpassed the average life span of the Babylonians, it was amazing they noticed the cycle at all.

Saroses are still with us, of course, and nowadays have numbers. Eclipse fanatics know them well and watch lovingly as each saros slowly evolves with time. A young saros offers eclipses that grow gratifyingly longer with every successive event. A dying saros, whose cycle started more than a thousand years ago, produces teaser totalities lasting just a few seconds, or eclipses barely grazing the poles that almost miss our planet entirely. After a few more eclipses, the saros ends, its number retired like that of a beloved pitcher.

Several different saroses are always under way, their clocks ticking. In most years, the world gets one total solar eclipse somewhere.

Consider that 3½-minute March 7, 1970, totality over Virginia Beach, which belongs to a young saros given the number 139.

This saros cranks out total (not partial) eclipses with paths that always move northeastward along the ground. In 1988, it presented its next event a third of the world west of Virginia—a 3¾-minute totality over Indonesia. Another eighteen years later, in March 2006, the same northeast-moving totality swept from Libya to Turkey. Its next return, another one-third of the world west, will provide totality to Cleveland and Buffalo in 2024—at least to those who don't draw the shades.

So now the stage is set for the next eclipses over our own neck of the cosmic woods. They are long overdue. The continental United States has suffered the longest eclipse drought in its history, a thirty-eight-year hiatus that will finally end on August 21, 2017. That's when saros 154 will carry the moon's shadow track southeastward from Oregon to Jackson, Wyoming, then over southern Illinois and Nashville to South Carolina. It will offer 2 minutes 40 seconds of totality.

After that, the next US eclipse will occur just seven years later—that homecoming saros 139 event in the sequence that started for me at Virginia Beach when I was a kid. Fifty-four years and one month, or one exeligmos, will have elapsed on April 8, 2024, enough time for the returning moon's track to move around the world and also shift slightly to the north, and for the length of totality to grow beyond four minutes.

That eclipse will appear longest over central Mexico. Then the moon's shadow will move northeastward like a twister, duly crossing over Tornado Alley from Texas to upper Ohio; swing directly over Buffalo and Rochester, New York, then over Burlington, Vermont; and continue on to bewilder the moose in northern Maine. En route, this path of darkness will cross the track of the 2017 eclipse to give a few thousand stay-at-homers in southernmost Illinois their second totality in seven years.

Two great eclipses await Americans between now and 2024. Will you go?

Not many do. In typical lecture audiences, only one or two in a crowd of several hundred raise their hands when asked who has ever seen a total solar eclipse. A few others respond more tentatively and say, "I *think* I saw one." This is like a woman saying, "I *think* I once gave birth." If you only think so, you've certainly never seen totality. What people are remembering is some long-ago partial eclipse. These are quite common, happen over every part of the world every few years, and always require eye protection.

The rarity of totalities normally makes them a pilgrimage destination. The only totalities over the mainland United States in the past sixty years were that Virginia Beach event, a mostly cloudy eclipse in Maine in 1963, and an eclipse visible from Helena, Montana, in 1979. Very few people are lucky enough to see one effortlessly from their own backyards.

If people had any real clue how astonishingly life altering the experience is, they would rearrange their lives to make any reasonably nearby eclipse a slam-bang definite. About ten years ago, as an aurora lecturer in Alaska for *Astronomy* magazine, I'd just led our group back inside into the warmth (it was March and –25°F) after viewing a riotous all-sky display of the northern lights (yet another Sun-induced spectacle). We were positioned directly under the shimmering auroral curtains, the pattern resembling what an awakening drunk might see after falling asleep beneath some draperies that were somehow glowing. I had done this for three winters in a row, but this particular aurora was the best I'd ever seen.

The guests hailed from many countries, and all were wealthy enough that this wasn't their first nature expedition to an exotic locale. They were surely a likely group of eclipse chasers. "How many of you have ever seen a solar totality?" I asked, and counted the hands. Sure enough, twenty-six.

"Allow me to take a survey," I said. "Which do you regard as more spectacular — the aurora you just saw or the total eclipse?"

All but one chose the eclipse.

This may seem odd. Everyone has seen photos of totality, with the familiar black cameo of the moon surrounded by the Sun's corona. That image takes no one's breath away. It's not in the same beauty league as the Grand Canyon or even Katherine Heigl. Many assume totality to be little more than the simple experience of blackness. Again, nothing special: you can easily reproduce darkness by not paying your electric bill. Why travel to God knows where? Indeed, since totalities from 2010 through 2016 are mostly over the ocean, the primary way to see them is via an expensive cruise. Broken down, a totality typically costs about $1,500 a minute, dwarfing all of life's other costly indulgences, except possibly an outer-space ticket aboard a Russian space shuttle or, sometimes, marrying on impulse.

But the fully eclipsed Sun is always a breathtaking surprise. No one is prepared for it. Cameras never capture the true visual appearance. The reason is simple enough: the inner corona is bright, the outer corona faint and delicate. The correct exposure for one part of the eclipsed Sun either underexposes the other so that it's invisible or overexposes the eclipse into a useless burnout. A real eclipse does not resemble the ones on TV nature documentaries.

Then you have the, shall we say, *vibe?* I hesitate to revert to hippie-speak, yet it's no secret that a powerful feeling accompanies some experiences, which makes them far transcend the merely visual. A birth, for example. On film, childbirth seems to be a painful, unpleasant mess, culminating in possessing a wrinkled baby with globs of goo or blood spatters and a slightly malformed head. I've always said "No thanks" when invited by close friends to be present at a birth. Yet when I attended the birth of my own daughter, it was sheer magic. Totality is the same. Something happens when the Sun, the moon, and your spot on Earth form a perfectly straight line in space. It almost knocks you backward. People weep. Animals fall silent or make peculiar sounds. Those old accounts — they're all true.

The magic really starts about ten minutes before totality, with the Sun partially blocked but almost gone. You still need eye protection at this point. I prefer shade No. 12 welding goggles, because they display a clearer, optical-quality image compared to cheap plastic eclipse glasses. (Get them from a welding supply store, which is absolutely never in a mall, but rather in the worst neighborhood of your town, commonly adjacent to a fenced-in yard guarded by snarling dogs.)

At this stage, the Sun looks like a weird crescent, but the most amazing sight, and the correct place to look, is the surrounding countryside. Colors are saturated; shadows are stark; contrast is ramped way up. Ordinary objects such as cars seem somehow unfamiliar, as if illuminated by a different kind of star. It's otherworldly.

Expectation fills the air. Breathing changes. A minute or two before totality, shimmering dark lines suddenly wiggle over all white surfaces. These are called *shadow bands,* and *they can't be photographed.* Go ahead and try. When you look at the images later, you'll see the surface, but no wavy bands at all. The rather anticlimactic reason is simply that shadow bands have extremely low contrast. Because they shimmer, the eye readily picks them out. But they lie below the contrast required for a photographic image.

Then comes totality. The brighter stars come out. The Sun's corona leaps across the sky, much farther than you expected. Its delicate, wispy structure, following the Sun's normally invisible magnetic field lines, depends on the part of the solar cycle you're in. With a glance, you'll know if you're at sunspot minimum or maximum. In the latter period, the corona is round and more symmetrical, as if the Sun's springs have been wound up tight, and all the power is being held in place, ready to pop. In the former, the quiet Sun paradoxically lets go with long, irregular coronal streamers. In either case, the glow is very obviously that of a different kind of light from anything you have ever beheld. No surprise: the Sun's corona is by far the hottest thing the human eye can observe. It's made of plasma, the fourth state of matter, rather than the atoms that comprise the solar surface and everything else around us.

The experience is almost too much. It does not seem of this life or this world.

Totality can last anywhere from one second to about seven minutes. It ends all too soon, yet it's more than enough to leave you addicted. Observers immediately start thinking about second mortgages, part-time jobs, or whatever it takes to place themselves under the moon's shadow again. It is the premier visual experience of everyone's life.

But the darkened Sun also has a figurative dark side. Frustrated eclipse chasers tell strange, harrowing tales that rival the heartbreaking accounts of what people go through to pursue a true love or accomplish a lifelong dream such as climbing the Himalayas. As an "eclipse astronomer," a lecturer at half a dozen totalities, I've heard hundreds of tales. I'll share a few here.

A total eclipse is a sacred experience. Something very off-Earth is going on, and it forces one to tread lightly. It's not a place for bravado or ego. Even the famous no-nonsense astronomer Bart Bok, who observed many totalities with scientific instruments, watched his final one by simply gazing up with dropped jaw and no equipment whatsoever. Too many observers later regret spending the precious seconds fiddling with light meters and f-stops.

Words are superfluous during totality. Here we stand in the most hallowed halls of nature; what have we to add? I tell all eclipse groups that I am theirs until before the eclipse, but I cannot be found, nor will I speak, during totality.

During the August 11, 1999, eclipse over the Black Sea, the crew ran an ultra-long microphone cord up to a crow's nest at the top of the ship, and from that dizzying perch, I spoke through the PA system until about twenty minutes before totality. Then I wished everyone a good eclipse and shut up.

An hour earlier, another large ship, apparently seeing ours at anchor and figuring "This must be the place," stopped cold a mere half mile away. We could hear someone blabbing through that ship's PA system all through the event. He never stopped. Happily, we couldn't make out any words. What was that lecturer thinking? Would he also have tried to narrate the big bang or the Last Supper?

At the July 11, 1991, totality over Baja, Mexico, one member of our two-hundred-person group boastfully (and pointlessly) set up his fancy telescope in front of our hotel each midday and finally loudly announced that he had rented a Jeep to drive the forty-five miles up the beach to gain an additional quarter minute of totality. Normally, grabbing an extra fifteen seconds would be a wonderful idea, except this happened to be the longest totality for the next century. At six and a half minutes, you really didn't need any more, and I told everyone that we had an ideal location right where we were.

This fellow was so off-putting that I suspect everyone was relieved that he'd be elsewhere. He talked one other guy into going with him. An hour after totality, they returned crestfallen. At their location alone, an overcast had abruptly moved in. They were the only ones who missed the eclipse. We all sincerely felt bad for them.

A decade earlier, totality was to sweep across Siberia, and a group of professional astronomers gladly accepted the Soviet government's offer of an ideal observation site on an island in Lake Baikal. Theoretically, the surrounding water would provide stabilizing conditions for the instruments, and the island was smack-dab on the centerline of the impending event.

On the morning of the eclipse, however, two American astronomers overslept. The hotel had not provided the requested wake-up call. They missed the only ferry to the island. Disheartened, they resolved to make the best of it and set up near their hotel, where the eclipse would also be total, if a few seconds shorter. As it happened, a single stationary cloud formed over the lake. No one had anticipated that the cooling created by the hour-long, steadily growing partial phase would lower the air temperature to the dew point. Nobody on the island saw a thing. Only the late sleepers observed the eclipse. Of all the eclipse ironies, this experience seems most like some sort of lesson. But what is it?

Then there's the story of one of the greatest fiascoes in eclipse-chasing history. The 1998 eclipse over the Atlantic would be under five minutes, but an enterprising company chartered a Concorde in a clever attempt to squeeze out a longer totality. By flying faster than sound in the same direction as the moon's 2,000 mph shadow, the supersonic airplane would effectively increase the length of the eclipse to eight minutes. In reality, eclipses cannot be optimally observed through small, plastic airplane windows. But an even greater problem lurked: people on only one side of the plane could view the Sun.

All the passengers were briefed on the scheduled choreography. The guests, seated four abreast, would take turns occupying the window seats, sort of like musical chairs. Announcements on the PA system would tell them when it was time for those at the windows to leap up and yield their seats to the passengers to their right, while the others would then slide over to await the next switch.

By all accounts, it was a disaster. The plane did not fully turn into position until many precious seconds after the totality was already under way. After paying a fortune for a ticket, the first people at the windows reasonably wanted their promised two minutes, which meant extending their turn beyond the PA announcement. When they didn't vacate their seats, their neighbors insisted, with urgency at first and then with insults. This tense procrastination got passed along to each subsequent group. The final group barely got to see anything at all.

The plane landed in utter silence. No champagne, just smoke emanating from everyone's ears.

A friend who has been to seven eclipses was clouded out of four of them. In many cases, the clouds approach only in the hour or even minutes before totality. (Maalox and tranquilizers are necessary additions to any eclipse first-aid kit.) The good news is that NASA offers detailed climate and satellite maps a year before each event, with guidance as to where the sunniest places along the track are statistically likely to be. For the August 11, 1999, European eclipse, many masochists went to Cornwall, England, or to northern France or Germany, even though the climatological prospects were gloomy. By contrast, wiser travelers headed for Romania or Turkey, with their traditionally sunny skies. Not all eclipses follow the statistical odds, but that one did. Nobody west of Munich saw a thing.

A pair of eclipses that ran counter to common sense were the San Diego totality of 1923 and the New York City eclipse of 1925.

Many universities could afford to send only one all-out eclipse expedition. Which location had the greater chances of experiencing sunny weather, Southern California in September or New York in January? Are you kidding? Yet on the appointed days, the Golden State was overcast, while New York enjoyed a supernaturally cloudless winter morning.

The totalities coming up for the United States on August 21, 2017, and April 8, 2024, merit an advance look at the long-term cloud patterns over the entire track. Search online using the keywords "NASA solar eclipse." Study the weather prospects, then position yourself accordingly. Having a car and a willingness to travel fast are good ideas, too, so that you can move as soon as the forecast gets firmed up the day before. This might be a bit of a bother, but the reward will be nothing less than the most amazing thing you have ever seen. Like the Aztecs, you will experience the birth of a strange new love affair with the Sun.

AS FOR THAT 1980 totality at the Konarak Sun Temple, the sky remained cloudless as our Indian-made Ambassador came to a stop in the sand. My Indian companions and my Italian wife and I pried ourselves out, amazed at how few people had come to this ancient site to watch the eclipse. Where were the crowds? A couple from the Midwest had spread a blanket on a small sand dune, with the temple's erotic frescoes arrayed a hundred yards in front of them to the north.

Although an eclipse track is a hundred miles wide and thousands of miles long, some eclipses draw visitors to a single location. In Egypt in 2006, for example, only one corner of that country would be touched by the moon's shadow, so in that place alone, the government set up a special five-acre observation zone, complete with a military patrol against the possibility of terrorists lured by the sudden influx of foreign visitors. By contrast, the ribbon of totality on August 21, 2017, will slash across the entire United States. There will be no single "best place" to see it. The same was true in India that day in 1980, which is why, perhaps, the Sun temple was so sparsely populated.

Two hours later, after day had turned to night and back again, and the moon's shadow had moved on, everyone was visibly stunned. The midwesterners looked like they had seen phantoms from another dimension; they spoke in whispers and with evident difficulty. When asked her opinion, a member of our own party summed up the event in a very unscientific way: "I have just been to the home of my soul."

CHAPTER 17

That's Entertainment

The Northern Lights have seen queer sights,
But the queerest they ever did see
Was that night on the marge of Lake Lebarge
I cremated Sam McGee.

—Robert W. Service, "The Cremation of Sam McGee," 1907

WALTER AND ANNIE Maunder had guessed right. Just over a century ago, they wrote that dark spots on the Sun where magnetism is weirdly intense might hurl particles at Earth, which would light up our skies with the reds and greens of the aurora borealis. These ideas were, as we've seen, totally ignored. Then, in the 1950s, visionary American physicist Eugene Parker insisted that the Sun emits a constant swarm of high-speed atom fragments that he called a solar wind. Their paths would be guided by the spiral shape of the overall solar magnetic field, a nautilus design resembling the water thrown from a twirling lawn sprinkler and now known as the *Parker spiral.* But he was initially ridiculed, too.

Yet even when the "wind" was accepted, one vague and ill-defined player remained elusive in the psychedelic auroral drama: the true structure of our own planet's magnetism. This final puzzle piece was discovered by the American astrophysicist James Van Allen using equipment that he designed and that flew on the first US satellite, *Explorer 1,* in January 1958. He showed that our *magnetosphere* (a term coined a year later by astrophysicist Thomas Gold) has belts that trap and hold charged particles from the Sun. The picture was complete at last.

We knew that the Sun brings life, death, climate change, and evolution to all our planet's life forms. Now we could finally explain how it also creates jaw-dropping spectacles of premium entertainment value in high definition.

Nothing outside of a birth or an IRS audit can produce such sobbing or reverential silence like a total solar eclipse or the fabled northern lights. These Sun-choreographed crowd pleasers have in common the inability to show up adequately in photos or videos. Sure, you can see videos of them on YouTube, but the actual experience is a very different ball of wax. In person, the northern lights remain mesmerizing for hours. There's just something about half the sky, and sometimes the entire heavens, exploding with squirming, twisting ribbons of greens and reds, like modern art gone berserk.

You've never seen the northern lights? Well, you can. The sunspot near-maximum and post-maximum years of 2012 to 2016 ought to feature almost nightly displays if you go to the right place.

Start at Fairbanks, Alaska. Every display of the lights is just a small segment of a huge, five-thousand-mile-long glowing doughnut of auroras that encircle the magnetic poles. Fairbanks happens to sit right beneath this fluorescent oval. What clanking garbage trucks and sirens are to Manhattanites (omnipresent) the northern lights are to Fairbanksans.

With a population of just thirty-five thousand, Fairbanks is hardly a major metropolis. Its skies don't resemble the washed-out monochromes over Buffalo or Oakland. Nonetheless, one can easily dodge even its minimal synthetic sky glow by heading ninety minutes east and booking a cabin at Chena Hot Springs. Now, sitting in a natural 84°F lake at night, in the glassed-in aurora hut, or on any of the surrounding hillsides reachable by snowmobile, you can watch as the sky explodes almost nonstop.

When to go? Not summer, that's for sure. Not unless you love mosquitoes. Not to mention there's no night, so the whole exercise would be pointless. You want to be there in March or September. That's when you get twelve hours of night and the temperatures are tolerable.

I brought my twelve-year old daughter to Alaska in 2003. Unlike the attorney-dictated no-risk activities offered at resorts in the Northeast, I found Alaska's attitude intoxicatingly sensible—and now that I'm thinking of it, Hawaii's, too.

On the Big Island of Hawaii, atop Mauna Kea, I spent time working at some of the world's greatest telescopes, planted at 14,000 feet in seeming defiance of the volcano god, Pele. While there, I was astonished to see snowboarders hurling themselves off the volcano's edge. They'd leap off just slightly below the entrance to our Keck telescope, located at around 13,500 feet, where the air is so thin your mind barely functions. Indeed, that altitude scrapes the gasping upper limit of where unpressurized aircraft can fly. Yet there were no fences. No Keep Out signs. The rule was singular and simple: "No lights permitted" (because of the nearby telescopes).

How on Earth could a government-run facility allow this kind of risk taking? Weren't those in charge afraid of a lawsuit filed by the first person who broke his back—which, from the looks of it, would likely occur within a minute of a snowfall?

Not at all. Their tacit rule was simple: you kill yourself, don't blame us. Hawaiian courts have no sympathy for those who do stupid things and whose families then try to point fingers elsewhere.

The same was now true here in Alaska. "You want to use a snowmobile?" asked the resort's manager. I assumed he was treating me extra-special because I was the lecturer for the aurora-watching *Astronomy* magazine tour group that had booked the resort for half the month, but I wasn't sure.

"And your daughter—does she want one, too?" In truth, neither Anjali nor I had ever been on a snowmobile. Reminded of the dictum "Never lie to your doctor," we confessed we were total beginners, but that hardly gave him pause. "Nothing to it. I'll give you a lesson."

Back home, this lesson would have lasted three hours, cost $135, and entailed the signing of numerous legal waivers absolving the resort of any liability for facilitating our dismemberment. Here it lasted ninety seconds, and then we were off, following him up, up, and crazily up some more, rising so steeply that both of us were genuinely frightened that we and the heavy machines would tumble over backward.

At the top, we came to a canvas hut where someone was waiting with hot chocolate. The air was colder than I had imagined, let alone experienced. We learned it was –31°F. On impulse, I took a cup of boiling water, stepped out the door, and hurled the liquid into the air. I'll never forget the sound, a chorus of little *crink-tinkle-cracks,* and then nothing but ice pellets landing on the snow. Holy Toledo.

We were fifteen hundred feet above Chena Hot Springs and its little airport, which had held its own terrors for me earlier in the day. I rented a plane similar to my own and flew to the tiny Arctic Circle community of Bettles and then around Mount McKinley, returning by following the unrelenting barrenness of the Alaska wilderness for hours. It dwarfed the rural regions of my native Northeast, making Maine look like Düsseldorf in comparison. At the end of the flight, on my final approach to Chena, I had to zigzag around a prominent hill that sticks up like an implausible video game obstacle, then land—on wheels, not skis—on a snow-covered runway. At the far end (actually, about the last third of it), people were strolling around, seemingly unconcerned about the approaching aircraft. It was all so casual, as if life and death were simply left to unfold informally here in the Alaska wilderness.

Now, as night got ever deeper, we looked down at the isolated cluster of lights that were the cabins on one side of the unlit runway, but they held our interest for mere seconds. Rather, it was the surrounding craggy, snow-covered peaks that riveted us. As the sky darkened, an aurora filled the heavens. Usually, auroras arrive closer to midnight, but the unseen Sun had apparently decided to explode early that night, and the sky was already aglow with shimmering green curtains. The unearthly fluorescence illuminated every mountain for a hundred miles in all directions. All of Alaska's snowy peaks were emerald.

A good display of the aurora can be life altering, like a solar totality. Happily, such displays are often seen much farther south as well. In dark regions away from city lights, the northern third of the United States and all of Canada are treated to positively amazing auroras several times a year during the peak periods of the eleven-year Schwabe cycle. The trick is figuring out when exactly to pull the trigger and visit those friends in Vermont.

Say it's sometime between 2012 and 2016, you're psyched to see the northern lights, you have a car, and you're ready to drive. If you bring along some good city takeout for the deprived Vermonters, they'll call it even. But how to know ahead of time when to go?

That's a bit tricky. A coronal mass ejection that is aimed our way is likely to produce great auroras two or three days later, and spacecraft are monitoring the Sun nonstop. But those ten billion tons of approaching atom fragments have their own magnetic field, and an aurora can happen only if that field's north–south polarity happens to be aligned opposite Earth's. We won't know that until it sweeps past the ACE (Advanced Composition Explorer) spacecraft that measures such things. But that unblinking sentinel is just 900,000 miles sunward of us, and thus gives us only an hour or two of warning—not enough time for you to get to Vermont, the Upper Peninsula, or Montana.

So you have to take a chance. First, go only if the moon is not near full and, preferably, not even out. You want a dark sky, so avoid the period from four days before the full moon to four days after. You also want clear and not hazy skies.

March is a characteristically good aurora month, because Earth is tilted straight up and down relative to the Sun, but the aurora can occur anytime. This University of Alaska website—http://www.gedds.alaska.edu/AuroraForecast/—will give you advance warning of an aurora, but unfortunately it also issues many false alarms. Still, it's about the best you can hope for. If you're technically oriented and want to impress absolutely everyone you know, go to the ACE spacecraft's real-time website, http://www.swpc.noaa.gov/ace/MAG_SWEPAM_24h.html.

An aurora often starts as a simple, unassuming glow. If you already live in the country, you're aware of how the sky normally looks. A strange new brightness toward the north means either your neighbors are throwing a party and haven't invited you, that phosphorus factory in the next city has exploded, or there's an aurora in progress.

Keep an eye on the glow. Check it out every fifteen minutes. It might fade, or it might expand wildly into one of the following very cool auroral forms:

Patches or blotches that come and go.
Rays or streaks emanating from below the horizon.
Arcs of light, like irregular pale rainbows.
Vertical streaks that seem part of a rustling curtain.

All of the above may move leisurely or flicker so rapidly that five totally new scenes materialize every second. The lights will probably be pale green but can, much more rarely, be a rich cherry red. They may be confined to the north, even the low northern sky, or extend straight up, or even occupy the whole sky from horizon to horizon. If you see this, phone everyone you know no matter the hour—a 2:00 AM aurora call may even be appreciated by your ex and her new spouse. (Then again, maybe not.) If you live in the country and have nature-oriented friends, you can organize an aurora phone chain, so that anyone who sees them sounds the alert. We did that in my community, and our "northern lights alert" worked well for a decade.

One more cool fact is that whatever modern-art design unfolds in the sky above you, its mirror image is appearing in the far south, maybe in New Zealand. The aurora australis, the southern lights, are much less often observed, because no town exists under the south pole's auroral oval. But researchers who have explored and photographed these lights, and have compared the images to auroras happening at the same moment in Alaska or Vermont, have found that, second by second, their displays precisely match the same pattern of lines, curtains, blotches, and arcs, but reversed as in a mirror.

It's all too weird. But what exactly is going on?

Oddly, many Alaskans still have archaic ideas about the aurora—such as that it's caused by sunlight reflecting from oceans on the day side of our planet. But the process is now well-known and always follows the same sequence, from the unseen H-bomb energy outburst to the ultimate mind-blowing IMAX performance. First, the Sun erupts with a mass ejection of material that can be aimed in any direction. If the event happens smack in the middle of the Sun (from our perspective), so that the material is blasted our way, or at an angle that is nonetheless guided by the Parker spiral, the swarm of subatomic bees takes one to four days to get here.

These particles have electrical charges, so they interact with our magnetic field, which channels most of them safely around Earth. But as they sweep past, they penetrate closest to the ground where our magnetic field dives vertically downward, around our magnetic poles. This interaction between electrically charged particles and our magnetism generates electricity in the thin, "barely there" upper atmosphere. The action happens sixty to two hundred miles up—higher than jetliners fly, but lower than the International Space Station orbits. This is serious current. We're talking about 5 million amps and 50,000 volts. You don't want to plug your laptop into this.

Sometimes the solar particles are swept past Earth but then accumulate to spring back toward the midnight side of our world. Such a "substorm" was detected by the orbiting spacecraft THEMIS in 2008 and showed scientists that particles can be boosted powerfully by the changing magnetic fields they encounter.

Once they arrive, the particles create intense electricity that excites the thin oxygen ten times higher than the highest cirrus clouds. This makes the electrons jump to larger orbits within their atoms, where they don't like to be at all. In a fraction of a second, the electrons return to their preferred locations, and as they do so photons of light materialize. (Throughout the universe, light is created only when an electron falls down closer to its atom's nucleus.) Since the electron "jump" is of a known distance, the color produced is predictable: a pale green at 557.7 nanometers, similar to the hue of new spring grass. Rarely, people see a stunning red at 630 nanometers, which happens to be the same color as those red laser pointers. Our eyes are very insensitive to deep red light at faint levels, so we see this only on those rare occasions when the red oxygen emission is super-intense. Uncommon lower-altitude auroras are red, too, produced by glowing nitrogen.

But talking about the aurora is like talking about a great symphony or explaining a joke that someone didn't get. You really have to experience it for yourself. And if you are determined, you will.

The most controversial thing about the aurora is the sound. I've interviewed several Alaskans, plus one respected no-nonsense businessman in rural upstate New York who hunts yearly in central Canada, who all emphatically insist they've heard loud hisses and snaps during vivid auroral displays. Yet researchers at the University of Alaska just as emphatically say they've never heard any such thing, nor have they recorded any sound with sensitive microphones they've set up for just this purpose.

Since the northern lights happen too high up for any potential sound to carry, not to mention the long delay time from that great distance and the inevitable attenuation, sound would seem impossible. Yet I absolutely believe those who claim to hear something. I suspect that some people can somehow sense the huge electrical fields flowing over the ground beneath major displays. Perhaps metal fillings in teeth or some other as-yet-unknown mechanism can pick up the radio noise and static accompanying the electromagnetic component of intense displays. This would also account for the sound track being simultaneous with the visuals.

IF YOU DON'T live out in the country in the northern half of the United States or Canada or in northern Europe, and if you cannot afford to book a trip to Alaska, destiny might take its sweet time before gifting you with a grand auroral display. Happily, the Sun offers consolation prizes. These gorgeous apparitions are much more common and, unlike the aurora, are independent of the solar heartbeat, the eleven-year Schwabe cycle. They cannot compete with a great display of the northern lights, but they are nonetheless stunningly beautiful. And you can see them this very year, probably this very month, if only you take a daily moment to look for them.

These sunlight phenomena are produced by one of two processes: either *diffraction,* where light waves interfere with other light waves, or *refraction,* where each color of light bends at a different angle as it passes between transparent substances with different densities—as from air into and out of a raindrop or ice crystal. Refraction unscrambles the Sun's mixed-together colors, its whiteness. It neatly places solar emissions side by side in the familiar ROYGBV pattern. These fountains of glory are offered as complimentary gifts to all who look.

Let me start with my favorite display, which, alas, happens to be somewhat rare. I search for the *circumzenithal arc* (CZA) every sunny day of my life, and I see it once or twice a year. It is not subtle. It looks like an intense upside-down rainbow. This fluorescent smile is composed of the most vivid spectral colors one can ever behold. The colors are so rich, they make a rainbow seem washed out by comparison. Oddly enough, a CZA happens not in a Sun shower, but against a blue sky or slightly wispy clouds. No gloom or rain required.

As the name implies, this is an arc, about one-third of a complete circle, perfectly surrounding the spot directly over your head, the zenith. How to find it? Whenever the Sun is fairly low (but not extremely low), maybe a third of the way up the sky, block out the Sun's glare with an outstretched hand. Look high above the Sun. The CZA will either be there or (most likely) it won't. Look for it every day, and you will eventually see it. It's so stunning that you will stop strangers on the street and point it out to them and will desist in this proselytizing only when brought in for evaluation against your will.

When the majority of people see any kind of spectral pattern, they automatically say "rainbow," even though there are many lovely refraction phenomena that produce those same hues. Nonetheless, a true rainbow is not to be belittled, especially since it's the most famous and easily recognized materialization of the Sun's emissions.

No exploration of sunlight can ignore this fairy-tale aspect of solar energy, this marriage of science and fantasy. Here are ten rainbow facts that may be new to you.

1. In most places, rainbows are seasonal. In much of the United States and Canada, they are periodically observed late on summer afternoons and are extremely uncommon in winter. Reason: they require a Sun shower. In many parts of the world, rain usually arrives with an overcast sky. A Sun shower demands individual rain clouds, with the falling rain illuminated by sunny openings, and these are often more prevalent in summer.

2. Rainbows can never appear when the Sun is more than halfway up the sky. In early summer, that means no rainbows

between 9:00 AM and 4:00 PM. And since Sun showers are rarely morning phenomena, rainbows are usually apparitions of late-summer afternoons. (Except in Hawaii, that is, where they're common early and late on almost any day of the year.)

3. The lower the Sun, the grander the rainbow. Think about it: haven't the highest, most glorious rainbows of your life happened within an hour or two of sunset?

4. The sky above the rainbow is always much darker than the sky contained within the arc.

5. The highest the top of a rainbow can ever be is 42 degrees—or very nearly halfway up the sky. Again, such a maximally large rainbow will happen only at sunrise or sunset, in which case the rainbow will be deficient in blue.

6. The ends of a rainbow terminate at the ground simply because rain stops falling there. But you'll see more than a half arc, maybe even a complete rainbow circle—an enormous 84 degrees in angular extent—if you're positioned to look down into a sunlit water spray, such as a waterfall.

7. It's not rare to see a second rainbow 9 degrees outside the primary bow. It is always fainter, and its colors are always reversed. That's because it's caused by a second reflection inside each water droplet. No third rainbow arc is ever seen, although a series of alternating green and pink bands sometimes creates an inner fringe just inside the primary bow. These bands have the tongue-twisting name *supernumerary arcs*.

8. The empty space between primary and secondary rainbows is strangely darker than the rest of the sky. This is *Alexander's dark band* (reminiscent of the old Irving Berlin song and later movie). Point out the phenomenon to a companion, and she'll look at you with renewed wonder, since this is truly arcane knowledge. If you happen to be going out with someone special, up the ante by explaining that it's named after Alexander of Aphrodisias,

who lived around AD 200. See if invoking his name has any salutary effect.

Rainbow insanity time: The preceding facts were interesting. The next two are mind-bending.

9. Every rainbow is an arc, meaning part of a circle. And every circle has a center. Well, what occupies the center of every rainbow? Think about it, and don't say a pot of gold. (That's located at the end of the rainbow.) Okay, ready? The precise middle of every rainbow is marked by the shadow of your head.

Rainbows have a fascinating geometry. They are colorful, 42-degree arcs surrounding the point precisely opposite the Sun from where you are standing. Since the Sun is up in the sky (duh!), the *antisolar point* is always below the opposite horizon. It is where your head's shadow is cast on the ground.

The best way to visualize the geometry is to picture a rainbow occurring on the surface of an imaginary cone whose point, or apex, is your eye. The cone is always aimed precisely away from the Sun, and the cone's spread, or angle, is 42 degrees. It doesn't matter where on this cone the sunlit raindrops appear. They can be miles off in the distance, or they can be on the close-in smaller part of the cone near your eye, as on the spray from a lawn sprinkler a yard away. Whether from droplets near or far, a rainbow always has the same apparent size.

10. I've saved the best for last. A rainbow, not being a real three-dimensional object, does not have a reflection. Nor can it cast a shadow. And it is seen only by you. The person next to you is at the apex of a separate cone with totally different water drop-

lets. He sees a separate rainbow. If you hold out a mirror and observe a rainbow in it, it will be the image of a *different* rainbow. You can never see a rainbow *and* its reflection.

So when nobody's watching, is the rainbow there? No, it is not. Your eyes (or your surrogate, a camera) are needed to complete the geometry. The triad of Sun, water droplets, and observer are all required for a rainbow. When no one is present, we can picture the situation as an infinity of *potential* rainbows, each slightly offset from the others with various color emphases (since bigger droplets produce more vivid rainbows but rob them of blue).

Moreover, only when neurons in the retina and brain are stimulated by light's invisible magnetic and electrical pulses do they conjure the subjective experience of spectral colors. For both reasons, *we* are as necessary for rainbows as the Sun and the rain.

NOW LET'S LEAVE the very rare and the somewhat rare and move on to the downright common. These are the alley cats of nature. We're talking halos and Sun dogs. You can see these once or twice a week. Whenever the sky has wispy, feathery cirrus clouds or is solidly blanketed by high, thin clouds, block out the Sun with an outstretched hand and look far around it. Bingo: a ring. The inside of the ring is always red (unlike a rainbow, where red is on the outside), and sometimes the rest of the ring is mostly white. This is the 22-degree halo. It is always the same size.

Stretch out your hand fully to arm's length, spread apart your thumb and pinkie finger as far as you can, and close one eye. Place your thumb against the Sun. Your pinkie will show you the halo's position. This works because, oddly enough, everyone's thumb-to-pinkie span marks off the same 22 degrees of sky when held at arm's length. If you have a smaller hand and fingers, you'll also have a shorter arm, which brings your fingers in closer to your eye. If this doesn't work for you, simple corrective surgery can fix it.

Often a bright white or vividly colored spot sits on that halo exactly to the right or left of the Sun, or in both places. This is known as a *parhelion* to science nerds. Country folk call it a *Sun dog.*

All you have to do to see halos and Sun dogs is look up when the sky has thin, wispy clouds. These phenomena are created by simple refraction (light bending) within the hexagonal crystals of those frequent, high-altitude clouds. That so few people see them means, sadly, that relatively few have acquired the daily habit of observing nature.

But some halos leap at the unsuspecting. These are the rings, or aureoles, commonly seen around streetlights on a misty night. Rings can also appear around bright lights such as the Sun or the moon due to simple eye irritation, such as after swimming in a chlorinated pool, or the not-so-benign intraocular disease glaucoma. Of far more importance than the intellectual "Is it there?" issue surrounding rainbows, we might urgently wish to know whether a colored ring encircling the Sun, the moon, or a streetlight is truly present, or whether it's instead produced only in our eyes. There's a fabulously quick way to find out. Just block out the light with an outstretched hand. If the halo is really there, it will remain, and indeed appear more prominent. But if it is caused by an eye problem, the halo will vanish the moment you obstruct the light source.

ONE MORE SUN phenomenon deserves mention. This is another common apparition, and it goes by the name *cloud iridescence.* It occurs on the fringes of white clouds and is much more easily seen when you're wearing sunglasses. Cloud iridescence is special because, unlike most of the other gorgeous apparitions just discussed, it is *not* caused by the bending and separation of the Sun's colors (refraction). Instead, iridescence comes from diffraction. Here, light waves interfere with one another, canceling out some colors while promoting others, so that brand-new colors are manufactured. The resulting hues are not spectral colors. They are not the waves that emanate from the Sun or from any star in the universe. They do not resemble the colors the Sun produces in its core, the violet, blue, green, yellow, orange, and red. Rather, they are novel creations, which is why staring at cloud fringes is not as crazy as it may sound.

These new colors are vivid purples, hot pinks, and aquas—the sort of psychedelic mother-of-pearl, peacock feather, or surface-of-oily-puddle hues that please the soul, or at least our hippie sense of aesthetics. They are nature's way of demonstrating how it can process the Sun's emissions to bake up something entirely new. And they are there perhaps on half of all sunny days.

The aurora alone depends on the solar heartbeat. It alone ebbs and flows with the Sun's internal pulse. It gets riotous and frequent at eleven-year intervals. These other spectacles are the Sun's steady gifts. They are capricious displays of its emissions interacting with our own world's suspended ice crystals and falling water droplets—none of which would exist if the Sun's energy didn't forever maintain Earth's hydrologic cycle.

In an age of quick-cut, fast-paced movies and MP3 downloads, some nature lovers among us still pause daily to be thrilled and inspired by the Sun's kaleidoscopic offerings. The curtain rises during that minute in which we step out of the car and walk toward the front door. This is when we can either look around the heavens or keep our eyes lowered to the ground. Grand in scale but tantalizingly ephemeral, this theater offers its performances in the very air.

CHAPTER 18

Cold Winds

There is no patent. Could you patent the Sun?

—Jonas Salk, 1954 [on being asked who owned the patent on his polio vaccine by journalist Edward R. Murrow]

WE STUDY THE Sun because it's the most awesome thing around. But we also observe it for selfish reasons. We want to know "What's in it for me?"—meaning, what will earthly conditions be like ten, thirty, even one hundred years hence for our children and grandchildren and the great-great-grandpuppies of my dog, Walnut.

A big part of this quest is practical, not academic: We need to know how seriously to tighten our belts and take steps to mitigate man-made global warming, or, as science calls it, *anthropogenic climate forcing,* which sounds way cooler. Should we triple our electric bills and make ourselves poorer by mandating expensive carbon-free energy production? *Blade Runner* windmills everywhere—but NIMBY? We'd all probably say "Absolutely!" if we knew it would save the planet. At the same time, we don't want to be suckers. We don't want to spend money if we don't have to. We've seen scientists be wrong before. They were all wrong about the universe's expansion fifteen years ago. And they were wrong about Y2K. Just because the vast majority of them are now sounding climate change alarms, should we believe them?

First, let's be clear about one fact: Earth's climate is a moving target. It changes all the time. It changes in cycles of millions of years and also in an up-and-down rhythm that usually lasts 200,000 years; it changes in cycles that repeat every few tens of thousands of years and also in cycles within these cycles that can persist for a century or two, or even only a few decades. There are shorter cycles, too, odd lingering times of recurring cold or rain lasting weeks or months, but then we use a different word: "weather."

When it comes to global temperatures, nothing new can happen—the only novelty would be a change that unfolds a hundred times faster than ever before. Even discounting the very early Earth more than 1.5 billion years ago, when our atmosphere lacked its current comfy oxygen levels, our planet has seen unimaginably diverse conditions. We've truly been there and done that. Some of these conditions stood at the very limits of what is possible. We were probably Snowball Earth 700 million years ago, a 10,000-century-long period of unbroken whiteness everywhere and worldwide temperatures of −40°F. This period, as well as the briefer glaciation ages that carried mile-thick ice through what is now Central Park, could not have been more extreme. (Ice can never get more than a mile thick because the pressure at the bottom then liquefies the ice. The same is true of mountains, except in their case the limit is 52,000 feet, not inconceivably greater than today's Mount Everest, at 29,035 feet.)

Historically, there have also been vastly longer periods of 130 million years or more when there was no ice anywhere, when everything was much warmer and sea levels were hundreds of feet higher than they are now. One hundred eighty million years ago, we were the single continent of Pangaea, and we endured a warm, wet global climate that didn't budge for millions of carefree raptor generations. In the midst of the Cretaceous age, the poles were up to 70°F balmier than they are now. Arctic T-shirts were de rigueur. The equator was an informal place with temperatures several degrees warmer than today and ties always optional.

In modern times, the collision of two tectonic plates pushed the Himalayas nearly six miles into the air, which blocked and altered large-scale wind patterns and turned the previously green north pole into a Mister Softee franchise.

Our climate today is unusually cold by historic standards. Human beings have never known a normal Earth climate, because things have been unusual for the past 2 million years, ever since our ancestors arrived. The current odd climate is an ice age called the Pliocene-Quaternary that started 2.5 million years ago. It's a time when vast, permanent ice sheets remain, as in Antarctica and Greenland. We're so used to it, we think it's normal. Within this present ice age, we get periods of more extensive glaciation (called *bummers*) that occur in rough cycles of 40,000 and 100,000 years.

The last glacial period peaked 20,000 years ago and ended about 11,000 years ago. We're now enjoying a respite, an interglacial interval called the Holocene. It began 7,000 years before the pyramids were built. According to the Milankovitch cycles we'll look at shortly, this current interglacial should last an unusually long time. It will probably preserve property values for at least another 50,000 years and could even last 130,000 years.

Most climatologists now think that humans, with our blundering mischief, can actually convert the present Holocene climate into something approaching the warmth of one of those bygone hot periods by radically altering the atmosphere. An Earth that's as ice-free as it has often been might seem benign, but getting there in a century or two rather than over many millennia changes the game. The current balance of plants and animals and germs and rainfall didn't just occur overnight; rapidly altering the norm would result in far more problems than benefits.

But if warming accelerates, there will be benefits. Global climate change will make some parched areas wetter, some freezing places comfortable, some sterile regions lush, and perhaps even some House subcommittee reports intelligible. It will encourage the growth of plants and trees, supplying them with extra rain and carbon dioxide (CO_2). There will be winners, such as eastern Canada and New England, and losers, such as Southern California and the Southwest, which will become good places to breed and race camels.

On balance, though, any rapid changes would produce more bad things, such as epidemics, than good things, such as an extended growing season. This is simply because opportunistic pathogens are programmed to prey on virginal organisms, which will lack the antibodies to them for centuries or millennia to come. Although the mass media focus on effects such as sea-level rise, the far greater threat will surely be innumerable plagues and blights in multiple levels of the biosphere.

But are we really headed that way?

To help answer that question, it's important to look at the different factors that affect our climate. Let's start with the most logical influence: the Sun.

Satellite and ground measurements show that when the Sun is high overhead, it delivers 1,365 watts of energy to the air atop each square meter of Earth. The air absorbs some of this, while the ground and clouds scatter and reflect about 38 percent of the remainder back into space. In theory, our surface should reach equilibrium with an average temperature not too far above zero. Yet this is not the case. Instead, we live on a pleasant planet with an average of 59°F. How come?

Air. Our security blanket. Sunlight gets in to heat the ground because air is wonderfully transparent in visible wavelengths.

The same is not true of infrared, or heat. Those long wavelengths fly unimpeded through the N_2, O_2, and argon that make up 99.9 percent of dry air. But gases that have two different atoms, or three or more atoms of any kind—such as water vapor (H_2O), carbon dioxide (CO_2), and methane (CH_4)—absorb infrared as it tries to head outward toward space. They then reradiate this heat in all directions, including back toward the ground. As a result, the surface gets warmer.

That's the greenhouse effect. It's why Venus's stifling CO_2 blanket, ninety times denser than our own, gives our sister world the hottest surface in the solar system and has turned it into a don't-miss vacation destination whose nearest rival is Guatemala City.

The early twentieth century marked the beginning of a warmer period on Earth, which coincided with a major solar brightening. From 1948 to around 1990, the Sun emitted a higher radiance than at any time in the past one thousand years. The mid- and late-twentieth-century sunspot cycles 18, 19, 21, and 22 were all stronger than any since the George Washington administration. But solar irradiance fell in the waning years of the twentieth century, and the rug got pulled out after the start of the third millennium. Sunspot cycle 23's maximum in 2000–2001 was extremely wimpy, and its minimum in 2006–2008—and fully extending into 2009, even after the new cycle 24 began—was amazingly low and prolonged. This was a historical low. For months at a time, we saw no sunspots at all, while the Sun's brightness dimmed to a level not observed since before World War I. Sunspot cycle 23 was the second longest on record, bested only by cycle 4, way back in 1790. Between cycle 23 and the weak onset of cycle 24, 801 totally sunspotless days were observed — something that had never been seen by any living researcher. You would have to go all the way back to the Maunder minimum, which began in 1645, before sunspot cycles were even recognized, to find anything that would rival this extraordinary recent solar behavior.

At more or less the same time, global temperatures stopped increasing as rapidly as before. This seemingly shows that the Sun is still powerfully influencing our climate. Such relatively short spurts of solar variability, however, cannot explain ice ages. It turns out that our climate upheavals do stem from changes in the Sun's received irradiance, but they have nothing to do with it physically varying its output.

Huh?

This revelation was spelled out by the twentieth-century Serbian astrophysicist Milutin Milankovitch, who was initially ignored, in keeping with the scientific community's traditional response to original thinking. It's now generally acknowledged that he created the best explanation for long-term climate change. The Milankovitch theory states that as our planet travels around the Sun, variations in three elements of Earth-Sun geometry combine to produce changes in the amount of solar energy that reaches Earth, especially at critical latitudes. The three cyclical elements are eccentricity, obliquity, and precession, all of which sound impossibly geeky but are actually simple.

Eccentricity is the shape of our path around the Sun. Right now, Earth orbits in very nearly a perfect circle, with just a 3.5 percent difference between our *perihelion* (when we're closest to the Sun, in early January, at 91.6 million miles away) and *aphelion* (when we're farthest from the Sun, every July 4 or so, at 94.8 million miles). Over the course of 95,000 years, our orbit gets more squashed, becoming an ellipse with a 13 percent difference in distance. Then it eventually goes back to being more circular again. Since the energy we feel from the Sun depends on the *square* of its distance, an eccentric orbit greatly modifies solar irradiance, so that perihelion is then a much warmer time of the year than aphelion. Perhaps surprisingly, this change in orbital eccentricity is the least influential factor in creating ice ages.

Far more powerful is the second element, *obliquity,* or the tilt of Earth's axis. This tilt alternates between 22.1 and 24.5 degrees in a 42,000-year period. You wouldn't think a paltry variation of 2.4 degrees would even be worth mentioning. Yet, when the angle of our planet's axis is greater, seasonal differences are more extreme, since each hemisphere annually tilts more directly into and then away from the Sun.

The final factor is *precession,* or the direction our planet's axis points. Like a top slowing down, Earth's spin axis makes a slow wobble over 25,800 years relative to the background stars, or, more germane to climate, over 21,000 years relative to the orientation of our shifting elliptical orbit. This means that in half that cycle, or about 11,000 years, we tilt opposite to our present orientation, so that the Northern Hemisphere has summer in January, just when Earth and the Sun are closest. The result: much hotter summers and much colder winters.

Milankovitch developed mathematical formulas that figured how all three cycles either partially cancel each other out or add to each other to radically change the intensity of sunlight at various latitudes. His formulas predicted vastly changing global climate in periods ranging from tens of thousands to a hundred thousand years. For example, at 65 degrees latitude (mid-Alaska), where ice sheets can either form or fail to do so, Milankovitch cycles show that solar insolation, or sunlight intensity, varies by an amazing 25 percent, from 450 watts to 550 watts per square meter. (This is always less insolation than in equatorial parts of our planet, because high latitudes are forever slanted away from the Sun and thus get weaker sunlight.)

The interweaving of the three orbital and axial parameters create periods of great Milankovitch variations lasting about 200,000 years, followed by times when the cycles tend to cancel each other out, so that insolation doesn't change very much. We are living in one of the latter periods. Indeed, for the next 20,000 years, areas at 65 degrees north latitude will get slightly increased Sun intensity, suggesting that some "global warming" may be in the cards for us northerners no matter what. Although the Milankovitch theory needs to be refined, a study examining deep-sea sediment cores, published in the journal *Science* in January 1968, found that Milankovitch cycles really do predict and match the dates of climate change and glacial periods. After that supporting article, Milankovitch's explanations went from fringe to mainstream almost overnight.

Milankovitch showed that the Sun's incoming energy is ultimately an influential force on Earth. But does the Sun rule in the shorter swings that dominate our individual lives? Does it rule in warming the planet today, or at least win the best supporting actor award?

IT IS HERE that we must return to that other little player in the world temperature tragicomedy: our atmosphere. Trapped air bubbles in deep Antarctic and Greenland ice core samples divulge more than just the history of earthly temperatures. They also reveal that over the past 780,000 years, CO_2 levels have marched precisely up and down, in perfect synchronicity, with periods of global heat and cold. CO_2 has fallen to 180 parts per million (ppm) during the coldest historic periods and risen as high as 280 ppm (but never higher) during the warmest. Thanks to the human fossil fuel–burning frenzy, we have managed to raise the air's CO_2 levels from the eighteenth-century natural background level of 272 ppm to the current 388 ppm. Each breath we now take has 30 percent more CO_2 than the ones Thomas Jefferson inhaled. That's a huge change. It's too bad we're not plants.

Since CO_2 is a greenhouse gas that traps and raises heat on every planet on which it's found, Earth doesn't get some sort of special exemption. This alone is more than enough to explain the global temperature rise. Indeed, this major CO_2 boost ought to raise temps quite a bit more than it has. It probably takes a while.

What's disconcerting is that the link between CO_2 and temperature has historically gone both ways, in terms of which causes the other. The two definitely go up and down together, but when Milankovitch cycles have brought extra solar heating to sensitive latitudes, which has forced ice ages to end, it was the temperature that rose first. After a delay of a century or more, the growing warmth triggered the release of CO_2 trapped in the oceans, the way heated sparkling water increases its bubbling. So heat came first, followed by a CO_2 rise. This is a classic *positive feedback loop,* because the extra CO_2 then created further temperature rises. It became a runaway, like a snowball rolling down a hill, which picks up more material, which in turn adds to the forward inertia: one process reinforces the other synergistically.

In the past, the warming-up process happened naturally. But nature doesn't care what starts the snowball rolling. What's bizarre is that some people, educated people, actually imagine that there will be no consequences for releasing eleven gazillion tons of smoke each nanosecond and that we possess some sort of "Get out of smog free" card. The poorly informed want to blame the

1.5°F global temperature rise over the past century on the Sun's increased intensity, and they are mostly correct—if they limit their history recital to pre-1975 or so.

Something very weird has gone on since then. Before 1975, Earth's temperatures marched up and down with solar changes. Irradiance was as low as 1,364 watts per square meter in the late seventeenth century, and sure enough temperatures averaged nearly 2°F lower then than they do today. Solar irradiance climbed during the early twentieth century, and so did global temperatures—almost in perfect lockstep. It looked like the Sun alone was responsible for our planet's changing conditions.

Global temperatures stayed perfectly stable and even fell a bit during the entire baby boom generation, from the end of World War II until the hippies turned thirty or so. Is it any wonder that no one worried about carbon?

Actually, one person did. In 1938, Guy Stewart Callendar wrote what might arguably be the most important article of the twentieth century, published in the sleepy *Quarterly Journal of the Royal Meteorological Society.* Its title gave no clue to the ultimate tempest it would produce: "The Artificial Production of Carbon Dioxide and Its Influence on Temperature."

Here, and for the rest of his life, that British engineer and scientist explained exactly what happens to the air thanks to the burning of coal and other fossil fuels. Callendar amassed data for the first three and a half decades of the twentieth century from weather stations around the world. These figures showed an unmistakable warming trend. He examined natural and man-made sources of carbon, how the carbon was absorbed, and how much remained in the air.

Callendar was the first to correctly estimate that the natural background level of CO_2 is 290 ppm (we now say 272), and he was the first to document a 10 percent CO_2 increase between 1900 and 1935. Since this also closely matched his estimate for global emissions from burning fossil fuels, he concluded that most of the carbon was staying aloft and little was being absorbed. "By fuel combustion man has added about 150,000 million tons of carbon dioxide to the air during the past half century," Callendar wrote, estimating that three-quarters of this had remained in the atmosphere.

That was his only real mistake. Others later erred, too, but in the other direction, by insisting that 85 percent of emitted CO_2 is taken in by the forests and especially the oceans. The idea that CO_2 would be absorbed naturally was a major reason carbon emissions did not alarm many people even into the 1970s. Until recently, the real figure for natural absorption has been about 50 percent, with the latest exact proportion at 43 percent—that is, 26 gigatons of CO_2 are now released annually, and 15 gigatons remain in the air. There is some truly frightening recent evidence that the oceans are approaching the limit of what they can easily absorb without turning into big pools of crab-flavored seltzer. So they will increasingly not be able to pick up the slack, leaving more and more CO_2 in the air.

With his sophisticated grasp of how heat is absorbed and reradiated by CO_2, Callendar established the modern theory of climate change in his 1938 article. One year later, he elaborated in slightly more apocalyptic tones: "As man is now changing the composition of the atmosphere at a rate which must be very exceptional on the geological time scale, it is natural to seek for the probable effects of such a change. From the best laboratory observations it appears that the principal result of increasing atmospheric carbon dioxide... would be a gradual increase in the mean temperature of the colder regions of the earth."

Was this guy a prophet, or was he simply smart? Just as he decreed, Earth's coldest regions have warmed the most. In early 2011, scientists found that the Arctic Ocean is changing so rapidly largely because ever warmer Atlantic waters are rushing there through the Fram Strait, near Greenland. Indeed, oceans began a steady temperature rise in the 1970s and account for the bulk of our planet's warming. Since water is eight hundred times denser than air, it is a far greater heat repository than our thin atmosphere.

Callendar estimated the increase in average temperature due to human addition of atmospheric CO_2 to be 0.01°F per year, which is pretty close to what's been observed over the past century. Earth's temperature has climbed 1.5°F since 1880. Of course, CO_2 emissions have exploded since Callendar's day, and so has the rate of global warming. Most climatologists now expect the next 1°F rise to take less than twenty-five years.

But temperatures peaked in the 1940s, as if to metaphorically match the inferno of the Second World War, then started falling. Global temperatures kept falling for the entire rock-and-roll and Beatles eras, right until bell-bottoms were out—for a whopping thirty years—even while CO_2 levels continued to climb. This was *not* what Callendar predicted.

Around 1975, the fun and games came to an end, as in Don McLean's "American Pie." Global temperatures started climbing, and now faster than ever before. If you've ever wondered why "greenhouse effect" did not become a mantra until around 1990, it's because that's when, for the first time, the Sun's output started getting more moderate but global temperatures continued to soar. The Sun's brightness fell during the wimpy sunspot cycle 20 around 1970, then plummeted through most of the even more anemic cycle 23 as the new century began.

It now seemed that the Sun could not be held accountable for the higher global heat, and the wake-up call was sounded.

Year after year, new temperature records were broken. Yet no one was prepared for 1998. That was more than the hottest year the world had seen since the invention of the thermometer, the warmest in recorded history. It was a spike that didn't even seem to belong on the same graph. To this day, when looking at a chart of global temperatures, one can identify 1998 at a glance—it's that jagged Everest at the top of the page.

This frightened everyone in the field. Six of the seven warmest years on record had occurred in the 1990s. At the same time, areas in the far north—exactly where theory assures us that carbon must manifest the strongest—started changing visibly. Alaskans could see vivid alterations in seasonal ice. Permafrost melted and houses built on it shifted. The tilting homes in Inuit communities started to look as if all the carpenters' levels had been suspiciously emptied of their spirits.

At Mohonk Preserve's respected Daniel Smiley Research Center in the Catskill Mountains near New Paltz, New York, where meticulous meteorological and natural records have been kept for 130 years, naturalist Paul Huth told me that the average first autumn frost now arrives eleven days later than it did in the nineteenth century. Birds that previously migrated now routinely winter over. Nature there is changing conspicuously.

Global warming isn't just an abstraction. People everywhere are observing it with their own eyes. I've seen its effects in Arctic communities and photographed retreating glaciers in Tierra del Fuego, one of the southernmost areas of the planet. No region has been immune.

The year 2000 brought in the millennial odometer change with the feeble solar maximum of sunspot cycle 23. And, of all the crazy developments, global temperatures started leveling off. When the smoke cleared, during the years 1999 to 2009, temperatures were almost flat. This so-called plateau—which included the very cold year of 2007—occurred at the same time as the most extraordinary solar minimum of the past two centuries.

Now, "plateau" can be misleading and has been wrongly interpreted by some. Although the *rate* of global warming slowed down during the opening decade of the twenty-first century, worldwide heat levels were already so high that new records managed to be set despite the so-called plateau. Indeed, 2010 and 2005 beat out 1998 in the "top three warmest years ever," according to the National Climatic Data Center—no small accomplishment.

Starting in 2006 and continuing right through 2009, the Sun went blank for months at a time, and its irradiance fell to its lowest levels since before World War I. Was that the reason for the anomalously cool year of 2007? Was the decreased rate of global warming of the 2000s due to a faint Sun? Is the Sun still calling the shots?

Many US conservatives and some industry spokesmammals jumped on the recent temperature plateau and crowed that global warming is a hoax or a conspiracy. No need to meddle, to impose rules, to curtail our freedom to hurl CO_2 toward the heavens. Even weather forecasters were climbing aboard the new "no global warming" bandwagon. A front-page story in the *New York Times* on March 30, 2010, described "tensions" between climate scientists and many meteorologists. The article reported that the former group "almost universally" endorsed the view that the Earth is warming and that "humans have contributed to climate change." But half the TV weather forecasters were skeptical that any global warming is occurring at all. More than a quarter agreed with the statement "Global warming is a scam."

Many people—especially picnickers soaked to the bone—traditionally regard weather forecasters as lying scum. But half of these forecasters have a degree in meteorology, so why does such a large minority disagree with climate researchers? Is it the long view versus the short? The answer is simple: most of these skeptics believe that the Sun was, is, and always will be the dominant agent of terrestrial temperature change. When our planet's warming rate stopped increasing just as the Sun went cold in the first decade of the brand-new millennium, it was the last straw. Surveys showed the public, too, becoming a bit more skeptical of global warming—right along party lines. Most Republicans disbelieve in any real climate or carbon problem, while most Democrats feel that our world is experiencing a crisis that requires intervention.

The clouds of angry smoke now surrounding this vital issue increasingly come from TV commentators and laypeople who have no training in atmospheric science or, probably, high school physics, and certainly not in statistics. They haven't a clue. News commentators aren't qualified to evaluate the science of this enormously complex global interplay of air, sea, and solar insolation. Yet we've reached the point where millions of people appear satisfied by the single set of "statistics" persuasively presented by advocates for one side or the other.

So what's the real story? Ninety-five percent of meteorologists worldwide (versus only 75 percent in the United States) believe in anthropogenic climate forcing. It would seem perilous to ignore their opinion. Would you board a jetliner if 95 percent of all aeronautical engineers distrusted its design?

BY THE SPRING OF 2010, I realized I was in over my head. I threw my hands in the air and knew I had to go to the National Oceanic and Atmospheric Administration's (NOAA) climate center in Boulder, Colorado, to find out what the top researchers think and why.

I arrived in mid-May, though snow had just fallen and the lawns were dotted with patches of white. A scientist came out to meet me, buttoning his jacket. If this was springtime in Colorado, global warming was nowhere in sight.

I spent the day on a whirlwind round of one-hour meetings with a cast of characters who had no doubts whatsoever about where our climate is headed and what blame the Sun deserves.

Rodney Viereck, a tall, lanky redhead whose office looked like it had just suffered a terrorist attack, offered me a chair that was missing its armrests and presented sharp metal stubs instead, while he spread out pages of data. The charts and numbers showed that as our planet is heating up, the air high above is cooling.

"If more heat is being trapped at the ground, it means less is rising," he said. "So the stratosphere is noticeably cooling. This is making our atmosphere denser and more compact." He paused to look at me, presumably to see if I grasped the import of what he was saying, and then elaborated: *"This effect can come from greenhouse gases, but not from the Sun."*

He opened up an astonishing graph that confirmed the homework I'd been doing over the past several months. It turns out we were confused for so long because we are simple animals at heart, always looking for a simple explanation for global temperature changes. The truth is that four factors affect our planet's heat swings.

First, yes, there is the Sun's changing irradiance, which can raise or lower temperatures at the rate of 0.4°F per decade. According to NASA research, from the low to the high point of a sunspot cycle, the change in energy transferred to Earth is roughly equal to the "greenhouse effect" from fifteen years of human CO_2 emissions at the 2010 rate. This is huge.

Second is volcanic dust. A major eruption such as that of Mount Pinatubo in 1991 blocks sunlight and lowers temperatures at the rate of 0.4°F per decade, although such effects actually last only a couple of years.

Third is El Niño, that great upwelling of deep Pacific Ocean water. When El Niño is strong, it raises global temperatures by that same 0.4°F per decade, although it, too, typically lasts only one to three years. Its partner effect, La Niña, has a smaller effect, but in reverse: it creates a bit of global cooling.

Finally, there is carbon dioxide. Its influence does not come and go. Rather, its levels before around 1975 were small enough that its effects were usually dwarfed and masked by the other factors. That's why temperatures during the first seventy-nine years of the twentieth century marched up and down with the Sun's changing irradiance while also displaying a quiet background climb. Only in the mid-1990s did CO_2 finally reach a level where it had a greater overall effect on climate than the Sun's variability.

Viereck and I stared at the graph. The solar and carbon lines crossed in 1994. The nonsolar carbon effect will get stronger with time and will be predictable. Unlike the other climate influences, greenhouse gases overwhelmingly act on the colder half of our planet, and the effects manifest as a rise in a region's minimum temperatures. Carbon sneaks in and does its work at night.

This is why nearly everyone in the media gets the story wrong. Record heat in June and July 2010 caused many commentators to suggest that perhaps climate change *was* responsible after all. But actual greenhouse gas global warming does not make summer days hotter. Rather, it makes winter nights not as cold. This being so, it tends to escape our awareness. It happens mostly while we're sleeping.

With all four factors graphed together, the global temperature line now wiggles up and down (but mostly up) with perfect logic. Suddenly, the huge spike of 1998 makes sense, for that was a major El Niño year and also a time near the sunspot maximum. Volcanoes were quiet. CO_2 was on the rise. Everything acted together to produce heat that year.

The flatter 2000s make sense, too. CO_2 kept rising, of course, but there were no El Niño years, plus the Sun's irradiance was at a historic low. No wonder the heat line's recent ascent looked more like the Appalachians than the Himalayas.

But it will not stay flat. Already, 2001–2010 was the warmest decade since records were first kept in 1880. The bizarre Schwabe cycle 23 minimum is over, and the oven has been relit.

The Sun's variable brightness, mitigated by ocean currents and volcanoes, dominated the global climate picture from the dawn of Earth until the mid-1990s. Only then did the Sun get relegated to second place, a new, lower status it will endure for the rest of our lives, and those of our children and great-grandchildren.

Anthropogenic climate forcing has now become the biggest player in global heating. If carbon emissions go unchecked, all indicators predict positive feedback loops: melting polar ice creates dark-water oceans that absorb more heat, which melts more ice and more permafrost, which releases more methane, and on it goes until the world is 6°F to 10°F warmer, mostly due to warmer winter lows at middle and high latitudes.

Climate change will then be irreversible, no matter what we do. Those are conditions our planet has not seen for three million years. The results will be spectacular. Rising sea levels will be the least of them. More prominent will be weather extremes, with violent unaccustomed paroxysms. Most prominent will be biological blights and diseases, as previously cold-hating pathogens spread to tasty new organisms in the plant and animal kingdoms.

LATER IN MY VISIT TO NOAA, I viewed a series of computer simulations that showed where global warming will have the worst effects. The western half of the United States and Canada were right up there with central Russia, becoming terribly dry and 10°F hotter seventy years hence. Since this "yearly average" heat boost will be experienced almost entirely as a rise in winter lows, it really means that those lows must increase 20°F to achieve a 10°F mean.

As the computer whirred, what really caught my eye was the eastern United States. This is where all the models have been insisting for the past decade that global warming will be minimal. Even by this century's end, the Northeast in particular will have ample rainfall and only a 4°F temperature boost. That means winter nighttime lows in New England will go from the present 12°F average to 20°F, while summer temperatures will scarcely budge from where they are today. I found myself thinking, *Gee, that doesn't sound bad at all. In fact, it seems great.*

If it weren't for all those biological changes.

I then went to a very special laboratory, led by one of the directors, Donald Mock, who dwarfed everyone else at NOAA. At six foot six, he had to bend down awkwardly, like Alice trying to get through one of those tiny portals, to unlock the door to the lab where the world's carbon dioxide is measured and monitored. Samples are taken twice a month from one hundred clean air sites around the globe, in addition to the main sampling centers at places like the south pole and atop Hawaii's Mauna Loa. Here in Boulder, NOAA sniffs the global air nonstop. They even pay a little old lady in Mongolia, who releases her automated twelve-foot NOAA air-tasting device every two weeks and then carries the canister three hundred miles to Ulaanbaatar. There's no FedEx there, so she takes it to the US embassy, where it's carried by diplomatic pouch back to the States and then on to this lab.

"We monitor eighteen atmospheric gases," Mock said. "If a city in China switches to a different kind of coal, we can tell."

He pointed to a chart, a graph of carbon dioxide, with its annual sawtooth pattern superimposed on its steady overall climb. When the Beatles broke up, global CO_2 was 320 ppm. Now it's nudging 390 ppm. The rate of climb is increasing, too. The CO_2 level will break the 400 ppm barrier around 2016.

I finally broke the silence. "We're screwed."

Mock's eyebrows, close to the ceiling, scrunched together as he looked down to make eye contact. "Yes," he said quietly, like someone who has given up. "We're screwed."

Yet despite this unanimity among climatologists, many eyes still turn to the Sun. First, solar astronomers disagree as to what it will do next. The prolonged deep minimum and delayed start of cycle 24 is a strong indication that this cycle will be very weak, culminating in an extremely low maximum sometime around 2013. That would presumably be followed by another low minimum. If these events materialize, anthropogenic climate forcing will be kept suppressed, and the global temperature curve may rise only slowly.

There exists a minority view that cycle 24 will be normal and robust. I'll spare you the technical justifications. If so, the maximum may be followed two years later (meaning 2015 or so) by an El Niño year, which will raise temperatures on its own. If this scenario develops, it will be 1998 and 2010 all over again and then some. Global temperatures will spike sharply and then continue their upward march.

In summary, although temperatures recently plateaued, they're still warmer than ever. And the situation is far from stable, as 2010 demonstrated. The sea has risen one inch since 2000. Three hundred thousand square miles of Arctic sea ice have melted since then, too. And what shall we experience as the Sun reaches its next maximum?

Most of the solar experts—not to mention *The Old Farmer's Almanac*—think the Sun's recent cooldown will continue. Citing strangely weak magnetic activity in the new sunspots now emerging, a paper published in the American Geophysical Union's journal *Eos* in 2009 attracted a lot of attention by predicting that the Sun might be entering an extended state of low activity. If this is true, the faint Sun of 2006–2009 will go through only a gentle

2013 maximum in cycle 24, followed by another deep minimum, which will effectively neutralize global warming's rate of rise for years to come, maybe into the early 2020s. If the Sun chooses this period for another lifelong Maunder minimum (unlikely, but who knows?), that would be the best thing it could possibly do.

Clearly, much uncertainty remains. If we want guarantees, we've come to the wrong solar system. No matter what, it will be fascinating to watch sunspot cycle 24, which was born so quietly in 2008 with a single high-latitude spot of the correct polarity, as it unfolds for what promises to be a pivotally important eleven-year life span.

One thing is for sure: by cranking down its irradiance, by producing sunspot cycles with lower peaks and deeper valleys, the Sun has lately come to our rescue. Its behavior has partially counteracted human fossil fuel emissions. The next maximum is almost upon us, and no one knows what will happen after that. The Sun is buying us time. Will we use it to start limiting the carbon we hurl into the sky? Or will the unscrupulous or myopic use the Sun's gift as false proof that anthropogenic climate forcing never existed at all?

CHAPTER 19

The Weather Outside Is Frightful

Sleepers, awaken
The night has gone and taken
Our darkest fears and left us here
And the Sun it shines so clear
And the Sun it shines so clear.

—Mike Heron, "Sleepers, Awake!," 1969

THE POLAR FLIGHT from Chicago to Beijing had been going normally when, suddenly, all radio communication vanished in a hail of static. The giant twin-engine Boeing 777 was all alone over the Arctic wasteland. The captain and first officer exchanged glances but said nothing; there was nothing they could do. Out here, out of range of the equator-orbiting communications satellites, high-frequency radio was the only link to the rest of the world, and now it was gone. Just as that happened, and unknown to everyone on board, nonstop radiation began penetrating their bodies. All the passengers and crew were effectively getting a dozen medical X-rays' worth of ionizing radiation.

MONTREAL'S VAST UNDERGROUND tunnel system was far from crowded that Sunday night after midnight, with only night-shift workers avoiding the cold (19°F) March night in the streets above. Suddenly, without so much as a warning flicker, everything went pitch-black. Battery backup spotlights illuminated wall sections like torches, and yet vast stretches were as inky as the catacombs of Paris. High above, elevators stopped with a jolt, their occupants trapped. Streetlights throughout the province of Quebec went out, as transformers blew one after another, plunging a quarter of Canada's population into a total blackout. By daybreak, three million homes were cold. The commerce, schools, and transportation system of Canada's second most populous province had shut down.

HIGHWAY ENGINEERS USING their ultraprecise GPS units failed to notice a little "WAAS lost" signal—which would have alerted them that their surveying equipment had degraded to an exactness good enough for commuter driving but not for engineering and surveying. The result was a road ultimately paved in the wrong place, which later had to be ripped up.

THE *GALAXY 15* SATELLITE, an engineering wonder sending high-def TV signals for cable providers, suddenly lost all commands from its controllers on the ground. Every other aspect of the orbiting machine was still operating, but its utter deafness to humans meant that it would soon drift useless, a total loss. Engineers on the ground were incredulous. It was barely five years old and was expected to last another dozen. What had gone wrong?

FARMERS IN KANSAS were in their planting runs, their machines automatically guided by GPS, which told them which row they were on and where to plant next. Suddenly, flashing lights informed them that the normally trustworthy ultra-exact satellite navigation had simply stopped. Unable to do anything more, they pulled off their fields and silenced their engines. They could not know that farming operations were destined to be delayed for thirty hours — an eternity in the ephemeral planting season.

THE ALASKA OIL pipeline inspector stared in disbelief. This section was only three years old, and here was deep corrosion. Oil would have to stop flowing, and this one-ton, three-foot metal section would have to be replaced. The cause was ultrahigh-voltage electricity. But how did such a current find its way here, in the middle of nowhere?

IN ALL THESE EVENTS, the cause was the same: space weather. At least that's what it's officially called. But the agent of destruction doesn't come here from Mars or the Orion nebula or even somehow from empty space. It always comes from the Sun.

Even when it's feeling sleepy, the Sun sends out streams of charged particles that were first theorized in 1958 and detected by a Russian spacecraft heading to Venus in 1962. This solar wind is composed of actual solid material, about a thousand particles per cubic inch, and its speed can be more than a thousand times greater than that of a high-powered bullet. Naturally, this affects Earth's magnetic field — and therefore satellites and any exposed objects in space.

On occasion, especially during the years around sunspot maximum, the Sun emits far more powerful bursts of particles that whoosh our way at more than a thousand miles per second. These are the now famous coronal mass ejections (CMEs) that were utterly unknown until the 1970s. CMEs pockmark our satellites and their solar panels like smallpox scars, and roughen the surfaces of planets and moons. They deform our magnetosphere so much that some of the particles penetrate our atmosphere, creating high-voltage electrical disturbances that generate currents along the ground, especially along power lines and pipelines. Their intensity—up to 20 million amps and 50,000 volts—utterly dwarfs the strongest lightning. In May 1998, a CME destroyed the *Galaxy 4* communications satellite, immobilizing 45 million pagers. How many emergency fire calls went unanswered that day? How many physicians received no emergency beeps? How many husbands failed to pick up a quart of milk?

The solar wind and CMEs get far denser and faster around solar maximum, and they were at their most destructive in 1989 and 1998, within a year of the solar maxima of their cycles. They have been unwittingly observed by *Homo whatever* for millions of years, since they push on comet tails like a gale on a weather vane, making them always visibly point away from the Sun.

Take that massive Canadian blackout. It started on Friday, March 10, 1989, when astronomers using solar telescopes (the modern fleet of Sun-monitoring spacecraft had yet to be launched) observed what they thought might be a CME.

We now realize that the Sun's complex magnetic field contains regions that can disconnect with a snap and then reconnect, unleashing the power of thousands of hydrogen bombs in a matter of minutes. This hurls billions of tons of disemboweled atoms into space.

The March 1989 explosion happened in the middle of the Sun's disk, facing us, so that the squirming swarm of homeless protons and electrons was aimed precisely in our direction. High-energy electromagnetic radiation was also unleashed, at the speed of light. It flew to our world in only eight and a half minutes and instantly disrupted radio broadcasts. The Berlin Wall was destined to fall later that year, but for now the cold war was still very much in progress and the CIA assumed the sudden jamming of Radio Free Europe into Russia was the work of Soviet electronics specialists. Instead, and recognized by no one, it was the first sign that we were under attack from the nearest star.

Two days later, on Sunday night, the enormous mass of solar particles struck our planet's magnetic field at a thousand miles per second. The cloud of charged particles had its own magnetism and, by a fifty-fifty chance, its polarity was aligned opposite our own—the only configuration that would allow it to transfer its energy to us rather than being guided harmlessly around Earth by the protective magnetosphere.

The solar detritus, channeled along our field lines, swarmed into our atmosphere near the poles. Instantly, it created enormous electrical currents in the high, thin air, which gave birth to the most spectacular displays of the northern lights seen in many years. You didn't need to be in Alaska to view them. They flickered brilliantly over me in upstate New York, from horizon to horizon. I activated our "northern lights alert" phone chain so that dozens of others could gaze at them, too. They were vivid even in the skies over Florida and the Caribbean.

While compasses went crazy, electrical currents began surging along the ground beneath the riotous displays. Garage doors throughout the continent started going up and down and continued all night long. The ground itself sizzled; areas with high concentrations of igneous rock, including most of Canada's eastern half, experienced geomagnetically induced currents in all power transmission lines.

At 2:44 AM on March 13, the Sun-induced surges started creating havoc in Quebec's power grid. The one-hundred-ton capacitor number 12 at the Chibougamau substation tripped and went off-line. Two seconds later, a second capacitor blew, and then one hundred miles away, at the Albanel and Nemiskau stations, four more capacitors went off-line. When yet another capacitor failed and five transmission lines from James Bay tripped, the entire 9,460-megawatt output from Hydro-Québec's La Grande Hydroelectric Complex was cut off. Within a minute, the Quebec power grid had collapsed. The province of Quebec was blacked out. Three million people were in darkness. More than half a million of them depended on electricity for heat. But nothing could be done. Dawn broke with no power whatsoever. The workweek began with the Montreal metro silent and useless. The city's main airport, Dorval, stripped of its radars, closed.

The storm came within a hairbreadth of engulfing the big country to the south. US electric grids were perilously close to a series of shutdowns. New York Power lost 150 megawatts the moment the Quebec power grid went down, while the New England Power Pool lost ten times that amount. Ninety-six New England utilities lost power but were able to borrow it from other reserves — thanks to the off-peak hour, when demand was minimal. A large step-up transformer failed at the Salem Nuclear Power Plant in New Jersey — one of about two hundred separate events in the United States, which included generators tripping out of service and voltage swings at major substations. And yet the country managed — just barely — to avoid a series of cascading blackouts.

Meanwhile, up in space, several satellites had their semiconductors fried by the intense solar bombardment. They tumbled out of control for hours before they could be brought back. The space shuttle *Discovery*, launched a few days earlier, suddenly started having problems, too. A sensor on one of the critical tanks supplying hydrogen to a fuel cell showed anomalous pressure. When the solar storm ended, the mysterious readings vanished.

That was little more than twenty years ago. It may seem like ancient history. Was that the worst that could happen?

Not even close. These days we rely more heavily than ever on electronics, transmission lines, and satellites. A decade ago, jetliners made a dozen polar flights; last year, there were seven thousand. But our brave new world has not included many brave new safeguards against space weather. For a peek at the perilous possibilities, let's rewind to August 28, 1859, when the Sun got unusually crowded with spots and flares that lasted for an entire week. A few minutes before noon on September 1, as the English astronomer Richard Carrington watched in amazement, one flare brightened so much that it doubled the Sun's brightness in that section of the solar disk. He could not, of course, see the X-rays streaming intensely from that spot; their brilliance would have dwarfed the visual emissions. Also unknown to him was the fact that this was the largest CME in the Sun's recorded history.

Normally, such a burst of material plows through the continuous, slower-moving solar wind, creating shock waves and gumming up the process a bit. But a CME the day before acted like an NFL blocker, clearing a straight, empty path between the Sun and Earth. Solar material usually requires three to four days to reach us. The CME on September 1 needed only eighteen hours to arrive at our planet, to this day the second-fastest journey ever recorded.

When this ultrahigh-speed piece of the Sun hit Earth on the night of September 1–2, 1859, the shotgun blast created the most powerful geomagnetic storm ever experienced. Back then, before power lines and pipelines existed, the only cables capable of carrying current were the telegraph wires that had been strung all over the United States and Europe just twenty years earlier. They started sizzling and popping in showers of sparks. The impossibly high current instantly found its way into populated areas. Sparking equipment plagued countless telegraph offices throughout both continents, as frightened operators leaped from their seats. Some did not move in time and were found unconscious on the floor.

The damage was extensive and ultimately costly. Meanwhile, brilliant auroras lit up the skies everywhere, even over Cuba. These northern lights went beyond the normal pale green; they contained vivid deep crimson at both low and high altitudes, alarming millions of people who had never seen nor imagined that the heavens could contort in such brilliant, twisted explosions. They were so bright over the western United States that people got up and made breakfast, assuming that dawn had arrived. In New York City, gawkers crowded rooftops and sidewalks.

The second-worst geomagnetic storm in recorded history occurred half a century later. According to a series of articles in the *New York Times* in May 1921, with headlines such as "Sunspot Aurora Paralyzes Wires," the entire signal and switching mechanism of the New York Central Railroad was knocked out of operation, and a fire raged in the control tower at Fifty-Seventh Street and Park Avenue, followed by a fire in the Central New England Railway station that destroyed the building. At the same time, telegraph operations throughout the country were disrupted.

Again, this was before electronics, pipelines, satellites, GPS, jetliners flying polar routes, a manned space station, and long-distance high-voltage power lines. Have things gotten less secure since then, or more?

I checked with NOAA's Space Weather Prediction Center,

which issued a detailed analysis following the relatively mild solar storm of November 2003. Outbursts of solar energy during late October and early November that year triggered severe geomagnetic storms with wide-ranging effects that included the following:

> "Strong geomagnetically induced currents (GIC) over Northern Europe [that] caused transformer problems and a system failure and subsequent blackout."
>
> "Radiation storm levels high enough to prompt NASA officials to issue a flight directive to the ISS [International Space Station] astronauts to take precautionary shelter."
>
> "Airlines took unprecedented actions in their high latitude routes to avoid the high radiation levels and communication blackout areas. Rerouted flights cost airlines $10,000 to $100,000 per flight."
>
> "Numerous anomalies were reported by deep space missions and by satellites at all orbits. GSFC [Goddard Space Flight Center] Space Science Mission Operations Team indicated that approximately 59 percent of the Earth and Space science missions were impacted. The storms...caused the loss of the $640 million ADEOS-2 spacecraft [which carried] the $150 million NASA SeaWinds instrument. Due to the variety and intensity of this solar activity outbreak, most industries vulnerable to space weather experienced... impact[s] to their operations."

Yet this was only a moderate solar burp. Unlike the 1859 maelstrom, this Halloween Storm, as space weather experts quickly called it, didn't even happen at solar maximum. What would an 1859-like event do to us in today's high-tech environment?

In May 2008, a US government workshop was held to answer that very question. For the first time, an all-star team of space weather experts was assembled under one roof. Their conclusions were anything but reassuring.

They estimated that a truly major Sun storm would produce damage of $1 trillion to $2 trillion during the first year alone and that this "severe geomagnetic storm scenario" would require a recovery time of four to ten years. Were they kidding? Why doesn't the public know about this threat? As one presenter at the conference noted with alarm, "Space weather events are a classic example of what social scientists call a low-frequency/high-consequence event—that is, an event that has the potential to have a significant social impact, but one that does not occur with the frequency or discernible regularity that forces society to develop plans for coping."

That 2008 report included a bottom-line CliffsNotes-like summary that can serve here as a good review of space weather basics. The report said that large-scale eruptions of plasma and magnetic fields from the Sun's corona—CMEs—can contain ten billion tons of coronal material and travel at speeds as high as 1,800 miles per second, or 1 percent of the speed of light. This amounts to a total kinetic energy wallop equal to the Sun's entire energy output for one second. The report noted that such events happen most often around solar maximum and are the result of energy released from the solar magnetic field.

The report said that solar flares and CMEs can occur independently of one another, but that both are generally observed at the start of a space weather event that leads to a large magnetic storm. Finally, to be maximally geoeffective and to drive a magnetic storm, a CME must be launched from near the center of the Sun onto a trajectory that will cause it to impact Earth's magnetic field. It must also be fast and massive, and have a strong magnetic field of its own, whose orientation is opposite that of Earth's.

Obviously, we're vulnerable. We cannot stop these things. So what are we doing about it?

The easiest way to find out was to meet with the relevant scientists. So I returned to Boulder, Colorado, where my press credentials earned me a couple of days nosing around with the remarkable characters at NOAA's Space Weather Prediction Center. The Sun was bright and seemingly benign the morning I arrived.

The offices and labs are located along hallways that twist and turn like a labyrinth. My escort navigated them with ease. I could only hope he would come back for me after each interview (he did), since I've never been great at the rat thing.

I asked researcher Doug Biesecker—a thin, energetic man in his early forties—who NOAA's clients are. "We have more and more corporate, private, and government customers whose livelihoods depend on being able to guard against solar storms," he explained. "It's big business."

We paused outside the TV studio–like glassed-in control center, where a series of monitors of the whole Earth and the Sun showed the latest information from various satellites.

"For airlines," he went on, "the primary issue is communication. They prefer to use high-frequency systems, as they are cheaper. At midlatitudes, space weather can cause periodic outages. In polar flights, outages are more severe. The issue for polar flights is that HF is the *only* communication method, so if it is out, they can't fly there, period. The secondary issue is health. There was an FAA advisory during the Halloween Storm of 2003 warning of radiation doses above 25K feet. Pilots lowered altitudes, and the cost was time and fuel."

I stared at the giant green Sun images from the twin *Stereo* spacecraft, which had been launched in 2006. One orbits outside

Earth's path and hence moves more slowly than our planet; the other stays within and travels faster. Both slowly change their viewing angles on the Sun.

"Communications are also an issue for emergency managers at the local, state, and federal levels," Biesecker continued. "And the Coast Guard. Pretty much anybody that is using high-frequency radio." He pressed a series of coded buttons to let us into the control complex.

"GPS is another big client. Lots of disparate uses here, from marine to aviation. Surveying, like for roads and building. Drilling. And precision agriculture. Planting and harvesting is now pretty much exclusively done with GPS, and when it's knocked out, as it was in 2003, it can cost a lot of money. And, of course, the satellite industry. Solar particles can induce a charge in their semiconductors, producing random commands. I can't even count how many satellites have been damaged or destroyed by solar storms."

We were inside now, and Biesecker lowered his voice. "Finally, the Department of Defense is always requesting our space weather information. The military uses the high-frequency FM band between 3 and 30 megahertz, and its signal varies one-hundredfold during the eleven-year sunspot cycle. There are many military applications. I can't talk about those."

"You've been doing this all your life? Do you still enjoy it?" I asked.

The room was very warm, and he wiped his balding redheaded brow. His eyes lit up as he said, "This is a dream job. I get to see what the Sun's doing every day!"

My next contact was Chris Balch, the lead space weather forecaster. I could see his gears turning as he quietly assessed my Sun knowledge. I could tell that, like most experts, he wanted to speak on the highest effective level I could grasp. But he began with a primer, patiently explaining the three types of solar storms.

"First, there are high-energy electromagnetic waves that come in at light speed. This includes extreme ultraviolet and X-rays. They ionize our atmosphere down to forty-five miles and can produce radio blackouts."

He pointed to a model of the Sun. "The second threat is radiation storms. We have neutron monitors installed in Maryland. We have threshold levels of what's dangerous, but anything approaching 100 million electron volts can be a hazard to airline passengers and extremely hazardous or even lethal to astronauts. These even make it all the way to the ground."

He tapped the right side of his Sun model, a crude high school science project kind of painted wooden ball that he apparently kept for this sort of interview. "We get more solar particle events from the Sun's western hemisphere. That's because the Sun twists its magnetic field, and particles from the right side of the Sun can more easily follow that field all the way to Earth."

The third and final category is the geomagnetic storm. This is the one that can cause electric grid collapses and power blackouts. Such a storm also sends currents of hundreds of amps along pipelines, damages transformers, destroys radio communications, knocks out satellites, and downgrades GPS capability.

The tool kit used by the space weather guardians includes old-fashioned magnetometers, like those in the days of Walter Maunder, and ground-based particle detectors. But the most valuable, not to mention the coolest, equipment is the satellites and spacecraft. In past decades, we've had *Ulysses* flying over the Sun's poles, and solar wind detectors on board the departing Voyagers. But the modern workhorses are GOES, SOHO, ACE, *Stereo,* and the new SDO.

GOES, one of the Geostationary Operational Environmental Satellites, monitors the Sun and Earth simultaneously from Earth orbit and measures the effects of solar storms once they've arrived. It looks at the Sun in the light of X-rays and has a detector to count the protons and other incoming bullets.

SOHO (Solar and Heliospheric Observatory), which is parked nearly a million miles sunward of Earth, stares at the Sun continuously and can see solar flares and CMEs with its coronagraph. It watches them being created, and thus usually provides two to four days' warning. SOHO, launched back in 1995 and now getting a bit long in the tooth, remains Doug Biesecker's favorite spacecraft.

ACE (Advanced Composition Explorer) is possibly the most valuable of them all, even if it has exceeded its shelf life and doesn't yet have a replacement. ACE, like SOHO, is parked at the Lagrangian L1 point, where the Sun's gravity balances our own. But it alone measures the density and magnetic polarity of the solar wind that sweeps past the spacecraft on its way to Earth. Hence, ACE gives us *specific* one to two hours' warning of a space weather event that will almost surely affect us. Even cooler for science nerds, ACE's real-time information is available to the public—meaning, you can see the same thing space weather experts observe. With a rare bit of happy familiarity the day I was there, I looked at ACE's recognizable EKG-type squiggles on one of the giant monitors in the control room.

Then there's *Stereo,* which is Chris Balch's favorite spacecraft—or, rather, pair of machines, since the *Stereo* twins continuously change their locations. *Stereo* observes flares and CMEs to perfection and shows dark coronal holes where the fastest solar wind blasts off unimpeded into space, to arrive on Earth a few days later. If a CME is violently propelled our way, it shows up as a symmetrical explosion: it *looks* like it's coming at us. *Stereo* also gets to peer around the Sun's back side, so it can see storms that are about to rotate into the hemisphere facing us.

A wonderful new satellite was launched in May 2010. All systems work perfectly on the exciting SDO (Solar Dynamics Observatory), and it is already returning stunning high-def images at ten wavelengths—including a visual band just for fun. Its job is to tie together the puzzles—to monitor the Sun's ever-changing magnetism, to stare at flares as they emit their brilliant X-ray flashes, to watch as pulses on the solar surface send sound waves through the interior, and to monitor the Sun in the extreme UV that can change its output a thousandfold in minutes, heating up and thickening our own atmosphere in the process. The SDO has enough fuel to function until 2020, and researchers are drooling like bulldogs at what it might uncover between now and then.

But researcher Juan Rodriguez (whose favorite spacecraft is GOES) warned me not to focus only on the glitzy, fast-paced period surrounding solar maximum, like the one in the new cycle 24 that should peak around 2013. "We've had more single-event satellite upsets during this recent *minimum* than ever before." He reminded me that cosmic rays from faraway parts of the galaxy or even other galaxies come in most strongly when the solar wind is quietest.

"Take a good video camcorder into a closet in pitch-blackness," he said, "and you'll see periodic flashes on the screen. These are impacts of cosmic rays. They are less frequent but more penetrating than solar particles, and we get more of them when the Sun is quiet. And unlike solar particles, these even reach the ground at the equator. So we can't win."

Just when the Sun blasts us most with pieces of its own body, it simultaneously uses itself as a shield to jealously block alien particles, making them arrive on Earth at only half their quiet-Sun rate.

I left NOAA's Space Weather Prediction Center, the Sun now sinking in the west, with a new appreciation for the people who think about our nearest star 24-7.

How far we've come from Wolf, Maunder, Schwabe, and Scheiner, not to mention Eratosthenes and his fellow Greeks. They had no less dedication or smarts than this twenty-first-century group. They just didn't have the tools.

And yet, for all the billions of dollars we spend, I was struck by how the Sun can still disable our fancy technologies on a whim, at any time, and in a way that takes the unaware public and mass media by utter surprise.

Arriving at my rental car and patting my pockets for the keys, I squinted up at what Shakespeare called the "flaming orb." It was good to know that the people in the building I had just left were always and unceasingly doing the same.

CHAPTER 20

Tomorrow's Sun

such a sky and such a sun
i never knew and neither did you

—e. e. cummings, 1944

TRYING TO MAKE things simple, some books say that stars are like people: "Just as we grow old and die, so will the Sun and stars." But this just doesn't work. We may be made out of the explosions of stars, but stars are not like us.

Will any person spend the final half of her life one million times smaller than she was before? Picture it—a woman going about her last thirty-five years the size of an apple seed.

And will any person's skin turn every color of the rainbow at various stages of his development, managing to skip only green? Will he blow out a smoke ring that hovers around him for fifty thousand years? Will his body go from a wispy near-vacuum gas to a solid almost infinitely harder than a diamond?

No, the Sun is not like us. We are sustained, modified, and influenced by it. We are made of Sun stuff. But it is an alien entity. Touchingly, we are like symbiotic organisms that watch their magnificent long-lived host with appreciation and also concern—because its death must automatically mean their own.

The Sun's fate does not rank among our leading neuroses. Many more superficial Armageddons make the evening news. Incoming comet. New virulent pandemic. Sea-level rise. Glue in Chinese toothpaste. Earth has been there and done that and survived it all. What we cannot survive is any significant change in the Sun itself.

Yet it is coming. It's already in progress. The Sun is becoming 10 percent more luminous every billion years. For its first 4.5 billion years, Earth endured the brightening light and adapted with a thicker atmosphere, more ozone, and inhabitants that learned to live with extra ultraviolet. Then, much later, humans made car air-conditioning standard rather than optional. But Earth cannot take much more. In just 1.1 billion years, a mere additional one-quarter of its already elapsed life, the Sun will radiate another 10 percent more energy. Climate scientists have tried to fudge their math and somehow make this luminosity go away. They've tried to discover some mechanism by which our planet can compensate. They cannot find any.

As surely as your taxes will go up next year and cats must inspect any empty cardboard box, the Sun's extra insolation will evaporate away the oceans. This will create a choking permanent cloud cover that traps even more heat on Earth. Global temperatures will stabilize at around 710°F—much hotter than your oven on "broil." Oh, well, at least it will be a dry heat.

Fleeing to the poles won't work. Life will have tried that. Perhaps some organisms will head underground, and that might do the trick, deep below.

But no complex life form has endured for a billion years, and neither will *Homo sapiens.* So we can speculate now as to how "we humans" will handle this distressing solar event, but in truth "we" will be unrecognizable from the way we are today. Our descendants, armed with at least twenty fingers thanks to keyboard evolution and hopefully a tail (helpful when trying to hold flashlights and wrenches—do you miss yours as much as I miss mine?), will perhaps follow the expected sci-fi route and flee to Mars, which will by then be quite a bit more balmy, even if the pizza is still lousy.

But that won't be the Sun's concern anymore. It birthed and sustained earthly life forms since it was just half a billion years old, and it must move on. The Sun and its creations will part destinies. The rest of this story—unless observed by our descendants from a more distant planet—involves the Sun alone, not us.

THE SUN'S LIFE as a normal G-class star, the kind astrophysicists call *main sequence,* is, at this point 1.1 billion years hence, still barely more than half over. For another four billion years, it shines only somewhat more brightly than it does today, still boasting the same size and radiating its customary creamy white light.

Forty million more centuries pass uneventfully. All this time, Earth is toast. The most soaring human accomplishments have long vanished without the slightest trace. If this sounds wistfully philosophical, well, who can be blamed for mourning the passing of this entire Earth experiment, with life forms too numerous ever to have been fully cataloged?

Unseen all these uncounted millennia, the hydrogen in the Sun's core is being consumed at an ever greater rate. But that's not enough to satisfy its new high-energy output. Increasingly, it eschews this simplest food and acquires a more sophisticated palate, converting helium to carbon and oxygen, just like its father before it. Now, four thousand million years after the death of Earth, as these processes become more dominant, the Sun starts emitting so much energy that its photosphere gets shoved outward. It expands enormously, its surface stretched all the way to Earth's orbit, those far-flung outer layers vacuum-thin, cool, and hence orange. The Sun is now a red giant.

Its width becomes a hundred times greater than before. Any human descendants on Mars, gazing through that world's thin air, see the Sun taking up a third of their sky. Between its gargantuan size and striking orange hue, it dominates the vistas of every planet like never before. It struts at its peacock best.

This new persona is neither stable nor long-lived. By now the Sun has endured for nearly ten billion years, but its red giant stage doesn't even last a single billion. With a final whoosh, its core performs its last trick, a desperate move to stay alive. The alchemic changing of its heavier elements to still heavier ones heats it up so dramatically and so suddenly that it explodes.

This is not a star-destroying catastrophe. It is not a supernova. This explosion merely blasts away the outermost 1 percent of its body, which rushes off as a bubble or ring. We see this happening to countless older Sun-type stars. It's so common that these distinctively symmetrical or ring-shaped gas clouds hover all around us in space. Observers in the late eighteenth century thought these strange, round, green blobs looked like the newly found planet Uranus, and so they called them *planetary nebulae.* Bad term. They have nothing to do with planets.

The Sun, starting to shrink, now heats up to its maximum and turns a fierce blue. It emits torrents of sizzling ultraviolet that catch up to and excite the expanding bubble it expelled and make its oxygen glow green and its hydrogen red. Here, ten billion years after the final episode of *The Simpsons* and fully eight billion years after the last reruns, our Sun is now the central star of a planetary nebula. Its blue surface seethes at an unbelievable 180,000°F—almost twenty times hotter than the 11,000°F of the long-gone human era. This Sun is as hot as any star in the galaxy. It is also collapsing due to its self-gravity. Its brief million centuries as a giant are over.

The glowing green bubble of gas that surrounds it keeps rushing outward, expanding, sporting red fringes like a holiday decoration. Its appearance probably matches the famous Ring Nebula, but it's not a duplicate, since the "original" Ring Nebula of our time has long since faded to oblivion.

The Sun's ring lasts just fifty thousand years, an *achoo* on the cosmic scale. The expanding gas then reaches a point in space too distant for the solar UV rays to excite it any longer, and the bubble fades to black. Anyway, the Sun's UV is itself weakening. The superhot stage of its life, like the red giant phase before it, has come to a close. This time, no new energy sources remain.

The Sun's nuclear furnace, out of fuel, sputters to an end. On its eleven billionth birthday, its pulse finally stops.

YOU MIGHT THINK the Sun would then blink off abruptly, like a tripped circuit breaker. However, energy is never lost in this universe, only converted, and the Sun still possesses *gravitational potential energy*. This means that any further contraction gets changed into kinetic energy, the energy of motion—in this case, atomic motion, which is another way of saying heat. So the collapsing Sun maintains its brilliance as it shrinks down and down, like Alice, its own gravity pulling it smaller, until, irony of ironies, it brakes to a halt when it becomes the size of Earth.

It is here that quantum mechanics comes into play and dominates the Sun's appearance for the first time. Electrons need wiggle room—they can't be too crowded—so the Sun's awesome in-pulling gravity is met by what physicists call *electron degeneracy pressure.* When the Sun has deflated like a slowly leaking balloon, crushing and packing its material until each sugar-cube morsel weighs a ton, it's simply too thick to collapse anymore. This new balancing point is super-stable. The Sun will remain the size of Earth forever.

Its torrid blue cools a bit until its body is uniformly white. It is now a white dwarf. It will shine this way for billions of years. We know this stage is common and long-lived because we see white dwarfs everywhere. They account for 6 percent of all stars. Indeed, every star weighing roughly what the Sun does ultimately collapses into a white dwarf. White means it's hot. Dwarf indicates Earth-size.

Time means nothing to this future Sun. No nuclear reactions occur. No pulse, no sunspot cycle, not even any spots. Instead, it sits unchanging, barely even spinning. But it is anything but boring.

First of all, having a Sun's worth of material crushed down to the size of Earth means it is 300,000 times more compressed than before. The Sun in our own BlackBerry era has an average density close to that of water. That's pretty thick for a gas, since water is fairly heavy, as you know if you've ever lugged a full aquarium from one room to another. That small aquarium, one foot on each side, contains sixty-two pounds of liquid.

But now the Sun implodes until its entire body is 125,000 times denser than steel. Each cupful outweighs a cement truck. It's much harder than a diamond all the way through. It's indestructible. You could hit it with an H-bomb without creating a dent. Compared to this future Sun, an iron cannonball is a vacuum.

But this Sun is not iron. It's an ultra-crushed ball of carbon, like a diamond, plus lots of oxygen. Of course, this isn't breathable oxygen; it's almost infinitely compact and in no way resembles a gas.

When we think *carbon and oxygen,* carbon dioxide springs to mind. That's by far the most common mixture of those two elements in our everyday experience. CO_2 is the notorious greenhouse gas of our atmosphere, although its actual concentration is just 0.0004, or a mere 0.04 percent of our air. The only place we routinely encounter it in far more concentrated form is when we exhale, where it is concentrated nearly one-hundred-fold. CO_2 is a significant component of each out-breath.

So the Sun has become one spherical, solid breath.

Billions upon billions of years now pass, the Sun shining mainly by its stored heat alone. Its white-hot surface gets further temperature replenishments not from fuel, for it has none, but from tiny contractions in size. The gravitational energy conversion is so powerful that when this white dwarf Sun collapses one inch smaller, that supplies ten thousand years' worth of white sunshine. But now its electrons have been squeezed as close together as possible, thanks to Wolfgang Pauli's exclusion principle, and there's not much wiggle room left for further contraction. This final power source comes to an end. Now it cools ever so slowly to become yellow, then orange, then red, eventually fading to brown and ultimately black.

Our Sun's true end is here, as a black dwarf. Not to be confused with a black hole. None of the imaginative black hole tricks are at its disposal. It cannot stop time. It's merely an ultradense ball the size of Earth. This dark, invisible entity is dangerous only because its enhanced surface gravity will remain a navigational threat to any passing spacecraft piloted by incompetent aliens.

But venture a million miles, and there's no danger at all. Its *overall* gravity hasn't changed much from that in our own era. Mercury and Venus have long ago been vaporized and Earth sterilized, but the other planets still orbit as before. The Sun's lengthy white dwarf period provided a mere point of light in their sky, a brilliant star that cast sharp-edged shadows but shone with far too little light or heat to be useful. Now, in the Sun's final, eternal black dwarf stage, the surviving planets appear to be pointlessly circling around nothing at all, like pathetic creatures of habit.

A time arrives when the dark Sun cools to exactly room temperature. Although it would now be comfortable enough, heatwise, you still couldn't set up a souvenir stand on its solid surface. You couldn't raise your hand. All its material has been gravitationally pushed down so forcefully that no part of its surface is even a millimeter higher than any other. It's smoother than a billiard ball. Living on this slippery perfect sphere, you couldn't get out of bed. The fierce gravity wouldn't even let you take a breath. Your body would be flattened, as if pressed by a steamroller. The Sun remains a "look but don't touch" entity even now.

This whole process, the Sun's entire biography—from its long-ago birth in the intermingling of two disparate nebulae to its ending as a cold, ebony globe—required twenty billion years. In our human era, not a single black dwarf is thought to yet exist. The cosmos just isn't old enough.

So the Sun's greatest adventures, its most astounding changes, happen only after it has destroyed all earthly life. It's almost as if it feels compelled to behave responsibly for the first half of its life before it can let itself run wild. Then, alone, it takes turns being enormous and tiny, and goes from wispy as a vacuum to crushed solid like super-steel. Then, alone, it changes color like a chameleon, from today's conservative white to orange to blue and back to white before going yellow, orange, red, brown, and then black.

We are left imagining something that has no reality anywhere in the cosmos now, in our own era of crickets and breezes. Having given and then taken life, the nearest star ultimately becomes an eternal black entity, unseen against the inky backdrop of space. Our beloved Sun then endures forever as an unchanging ball of frozen breath—a monument to its most curious creation so long ago.

Acknowledgments

I would like to thank the many solar researchers at the National Oceanic and Atmospheric Administration's Space Weather Prediction Center in Boulder, Colorado, for their generous time and support, particularly Doug Biesecker, and also physicist and historian Spencer R. Weart, for their valuable suggestions. Thanks also to John Cannell, MD, for his time with me and for his impressive dedication to fighting childhood diseases linked with the Sun. A special thanks goes to my editor, John Parsley, whose work on this book went above and beyond, and whose keen eye shines unsung through these pages. A heartfelt salute also to copyeditors Barbara Jatkola and Peggy Freudenthal. The book would have been clunkier, and sprinkled with more than a couple of errors, without them.

APPENDIX

The Sun's Basics

Mass: The Sun weighs the same as 333,000 planet Earths. Considering it's mostly hydrogen, the lightest possible material, that's an awful lot of substance. Two octillion (2 followed by 27 zeros) tons.

Mass loss: Like a diet run amok, the Sun loses 4 million tons of itself every second. Surprisingly, this won't visibly effect it for another 5 billion years.

Size: The Sun's diameter is 865,000 miles, the same as 109 planet Earths in a row.

Volume: 1.3 million Earths could fit inside the Sun if it were hollow. If we dropped copies of our planet into the Sun like a leaky faucet at the rate of one Earth per second, we'd fill the Sun after 11½ days.

Age: 4.6 billion years, same as Earth.

Spin: Like the moon, the Sun's average rotation is 27 days.

Surface temperature: 11,000°F, or 6,000°C. This is just barely hotter than the boiling point of tungsten. There could almost be pools of liquid tungsten there, but not quite. Just 100 miles above its surface, the temperature falls drastically enough to allow for clouds of steam.

Core temperature: 27 million degrees Fahrenheit, or 15 million degrees Celsius or Kelvin. It's essentially the same as a hydrogen bomb. No one can picture this.

Composition: Mostly hydrogen. For every 1 million atoms of

it, there are 98,000 helium atoms, 850 oxygen, 400 carbon, 120 neon, 100 nitrogen, 47 iron, and about 100 combined riffraff atoms of other metals. All as gas, of course. Since helium is four times heavier than hydrogen, the Sun's composition "by mass" is quite different, with 70 percent hydrogen, 28 percent helium, and 2 percent heavier elements, mostly oxygen and carbon. It is this last 2 percent that puts the Sun in the category of a "metal-rich" star. (To astronomers, everything except hydrogen and helium is a metal.)

Motion through space: 144 miles per second, toward the star Deneb.

Brightness: Magnitude –27, or the same as 450,000 full moons.

Notes

INTRODUCTION

The winner in the celestial short-name competition? Only Jupiter's moon Io beats out Sol, or Sun.

CHAPTER 1: YON FLAMING ORB

As an example of light that does not feel familiar and comfortable to all people, consider fluorescent lamps or the blue-white mercury vapor streetlights. These mostly emit narrow bands of violet, green, and mustard-colored light and have an "alien" quality, since they resemble nothing found in nature. Although their emissions mix in our eyes to produce "white light," people can sense that it is inherently different from the full rainbow spectrum that creates the white light of the moon, Sun, stars, or even incandescent bulbs.

The fusing together of hydrogen nuclei is the *main process* that makes the Sun shine. A small percentage of the Sun's energy output, some 1.7 percent, comes from fusing its helium into even heavier elements, a process that will become more dominant in the Sun's old age.

CHAPTER 2: GENESIS

If you enjoy a little math, here are the energy production figures for a typical star like the Sun. Each second, in the star's core, four million tons of mass is converted to energy. The process uses the famous formula $E = mc^2$, where E is the released energy, expressed in ergs; m is the mass being converted, expressed in grams; and c is the speed of light, expressed in centimeters per second.

We can work backward by first seeing what Einstein meant when he put the c^2 in the equation. Light travels 186,282.4 miles per second, so figuring its speed in centimeters per second is like figuring a horse's food consumption in oat flakes per decade. Yet this is what Einstein wanted us to do. Since there are 160,934.4 centimeters in a mile, a photon of light travels very nearly 30 billion centimeters per second. Then we must square it—multiply it by itself. This result—the speed of light expressed in centimeters and then squared—naturally never varies, so that the famous formula $E = mc^2$ could really have been written $E = m \times 900$ quintillion right from the beginning. Or, if you prefer, $E = m \times 900{,}000{,}000{,}000{,}000{,}000{,}000$. Already we're seeing that huge amounts of ergs of energy (that first E) lurk in each little gram of mass (m). That's why a few pounds of plutonium can destroy a city.

Now we simply multiply that 900 quintillion by the weight in grams of the material being converted. In the Sun, 4 million tons get converted each second, which, once we've expressed that in grams (a million grams make up a metric ton) and multiplied by the c^2 value of 900 quintillion, gives us the final answer: the Sun emits 4 followed by 33 zeros ergs of energy each second. By the time it spreads out and arrives here on Earth, it delivers 1,365 watts (365 more watts than a kilowatt, the average hourly home consumption of electricity) to the air above each square meter of Earth's tropics, where the Sun is high overhead. This is called the *solar constant.*

For more than a century, students memorized the blue-to-red star-family sequence with the mnemonic "Oh, be a fine girl, kiss me." When women increasingly enrolled in graduate astrophysics programs starting in the 1960s, the "girl" often became "guy." Then, perhaps influenced by campus pot consumption, the first few words changed to "Only beavers and fish...." What the mnemonic has become today, I cannot even guess. Yet the OBAFGKM letter sequence remains unchanged. Tell even an amateur stargazer that the Seven Sisters consists of type B stars, and she'll understand: "Ah yes, blue, young, hot, and massive."

What are the brightest explosions in the universe? There appears to be a variety of supernova that is one hundred times more powerful than normal. Some have called this a *hypernova*. Although it is not yet well understood, it may involve ultra-massive stars whose off-the-scale energy output creates antimatter-matter pairs, which then annihilate each other on contact.

CHAPTER 3: A STRANGE HISTORY OF SEEING SPOTS

Have you suppressed the Pythagorean theorem like so many others? It is simply that, in any right triangle, the square of the hypotenuse always equals the sum of the squares of the two other sides.

The Stoics were a school of philosophy founded by Zeno of Citium (ca. 335–263 BC). They gained such later famous adherents as the Roman statesman Seneca (ca. 4 BC–AD 65). But when nobody was watching, Stoics were observed to curse and yell whenever they hit their thumbs with a hammer.

Syene or Swenet, or Aswan as it's currently known, actually sits some forty miles north of the Tropic of Cancer, thus slightly missing it. On the summer solstice, the Sun, therefore, hovers one Sun width (0.5 degree) from the precise zenith there. Thus, contrary to myth, the Sun is never exactly straight up as seen from Aswan, but apparently close enough to shine down wells. In case you ever want to go there (it's a pretty place), be advised that it's one of the driest places on the planet: five years commonly pass between rainfalls.

Who invented the telescope? It might have been the Dutch spectacle maker Hans Lippershey, but it's still unclear. Several people claimed a patent, which, due to the controversy, was granted to no one.

CHAPTER 5: THE UNIT

If you're wondering how Horrocks's transit could have happened in November, when Venus transits occur only in June and December, it's because this was the old Julian calendar.

The plaque in Tahiti that celebrates the 1769 transit of Venus further said that the memorial "was restored and fenced round by the local administration at Tahiti and this plate was placed here by the Royal Society and the Royal Geographical Society in 1901." I have since learned that although Captain Cook's monument to the transit still stands, the plaque has disappeared. Having lasted a century, it is no doubt in the hands of an unscrupulous collector. (It wasn't me.)

Not many people have heard John Philip Sousa's "Transit of Venus March." Lost for a century, it was rediscovered by a Library of Congress employee in 2003, just in time to be played again to officially commemorate the 2004 transit.

Yes, Dan Blonsky correctly answered that million-dollar question.

CHAPTER 6: MAGNETIC ATTRACTION

Here is an interesting oddity. Mercury's trip around the Sun (88 Earth days) does not seem numerically linked with the Venus year of 225 days, so one would expect no correspondences between them. Yet we see thirteen Mercury transits across the Sun's disk per century—and thirteen Venus transits per millennium!

Alexander von Humboldt goes largely underappreciated today, despite the famous current named for him. The Humboldt Current is the largest cold, upwelling "river" of ocean water in the world. Flowing northward along nearly the entire west coast of South America, its low salinity and high oxygen content yield the most abundant marine ecosystem on the planet. But there's more: it was Humboldt who first proposed that the continents on both sides of the Atlantic fit together like a jigsaw puzzle, and thus must have been connected in the distant past. This correct idea preceded Alfred Wegener's theory of continental drift by a full half century.

Basaltic rocks reveal that our planet's magnetic field reverses its north-south orientation a few times each million years. This happens irregularly, rather than according to any sort of predictable timetable. Quite a bit of New Age literature has made people fear that the next flip of our poles is imminent and that this will cause widespread catastrophe. In truth, there's no reason that a magnetic reversal should occur during this century. Moreover, magnetic reversals have never matched times of mass extinctions, so they obviously do not hurt us much. Scout groups might wander around slightly lost, accompanied by disoriented pigeons overhead.

The exact nature of the Sun's rotation keeps getting tweaked. In September 2010, astronomers announced that they had found indirect evidence that the Sun's core, its center, which is the energy-generating innermost third of the radiative zone, may spin a bit faster than the rest of the radiative zone. Then comes the tachocline boundary layer, outside of which it's almost "every latitude for itself!" Outside the tachocline, the Sun's equator spins some 30 percent faster than the higher-latitude regions.

CHAPTER 7: THE WILD SCIENCE OF THE BEARDED MEN

William Herschel frequently expressed his belief that the Sun was inhabited. The Herschel quote at the beginning of this chapter appears in Edward Polehampton and John Mason Good's *The Gallery of Nature and Art* (1818).

In the yellow section of the Sun's spectrum, we see two black lines with the same position and spacing as the pair of yellow lines given off by laboratory sodium. Kirchhoff and Bunsen figured out why laboratory sodium shows bright yellow emission lines, while the Sun displays black lines blocking the same yellow part of the spectrum—almost like a photographic negative. When glowing, a hot solid, a liquid, or a high-pressure gas such as the Sun's surface emits all the rainbow colors, a so-called continuous spectrum. If, en route to us, this light passes through a cooler transparent layer made of, say, sodium vapor, the vapor absorbs sodium's fingerprint from the rainbow. This creates black lines, or gaps, in the rainbow with the identical pattern as laboratory sodium, but now black instead of yellow. Fraunhofer lines quickly started to be called *absorption lines.*

A spectrograph is simply a spectroscope that records or photographs its data.

Helium is the second most abundant element in the Sun, and in the cosmos, and one of the strangest. Helium bonds with nothing, which is why there is less helium in your body than uranium. Helium's normal form also makes it the only element without a natural solid state. No matter how cold, it never freezes. Rather, it forms a liquid with zero viscosity, meaning no slipperiness whatsoever. It runs right up the side of its container and down the other side, flowing away like an escaping eel.

Gustav Spörer remains one of history's unsung solar pioneers. He was not only the first to discover both major periods of Sun weirdness and sunspot absence in the fifteenth and seventeenth centuries, but he is also now credited with discovering that the Sun's poles rotate 30 percent more slowly than its equator. Although Scheiner, some 250 years earlier, noted that sunspots move at different speeds at various latitudes, he thought they were objects unattached to the Sun's pure, unblemished surface.

Maunder probably would not be remembered at all in our time were it not for the popularization of him (which amounted to near canonization) by pioneering twentieth-century solar astronomer Jack Eddy.

CHAPTER 8: CAUTIONARY TALES

The quote from Chaucer's "The Canon's Yeoman's Tale" comes from Larry D. Benson's *The Riverside Chaucer,* 3rd ed. (Oxford: Oxford University Press, 1988).

Two Israeli scientists, Lev Pustilnik and Gregory Yom Din, analyzed records of the price of wheat in England for nearly half a millennium, from 1259 to 1702. They found that Herschel was right: the cost of wheat was high during periods when sunspots were scarce or absent and low during solar maxima.

The supposed connection between the full moon and human births has provided a free-for-all, anything-goes setting for countless TV pseudo-documentaries and New Age magazine articles. Happily, numerous statistical studies have been performed, and the major ones are summarized in *Psychological Reports* 65 (1989): 923–34. The bottom line: human births are random as far as the moon's phases are concerned.

CHAPTER 9: WHY JACK LOVED CARBON

The Jack Eddy quotes in this chapter are excerpted from a recorded interview he gave to solar researcher Spencer Weart on April 21, 1999. They are reproduced here with Weart's permission and that of the American Institute of Physics and the Niels Bohr Library & Archives, American Institute of Physics, College Park, Maryland.

CHAPTER 10: TALES OF THE INVISIBLE

Most people use "heat" and "infrared" interchangeably, and that's usually okay. We do perceive infrared waves, whether from the Sun or from a heat lamp, as heat. And conversely, infrared detectors create images from the heat being emitted from an object. However, "heat" also can mean the speed that atoms are moving. If the atoms or molecules in a frying pan start jiggling faster, the metal is hotter. Indeed, "hot" and "cold" simply refer to the movement of atoms, period. When you run a fever, your body's atoms are jiggling a few miles an hour faster than before. This heat can be convected, or passed right through an object, always in the direction of heat moving toward cold, or faster-moving atoms speeding up those that are moving more slowly. So if metal has faster-moving atoms than those in your fingers (that is, it's hotter), contact will start your skin's atoms moving faster. If your atoms are made to move too fast, you'll yelp.

In 1782, the year after his astounding discovery of Uranus, William Herschel said, "Among opticians and astronomers nothing now is talked of but what they call my great discoveries. Alas! This shows how far they are behind, when such trifles as I have seen and done are called 'great.' Let me but get at it again! I will make such telescopes, and see such things!" And he did.

So what is there, hidden out of sight, surrounding the four-million-solar-mass black hole that lurks at the center of our galaxy? Mostly stars like the Sun. One star, named S2, orbits the black hole every 15.2 years; its motion lets astronomers probe the physical properties of the black hole. There are also hot, dusty gas clouds. They all emit copious infrared energy.

Heat, or infrared, along with radio waves, cannot ionize atoms or damage genes, and this is why both have long been regarded as benign as far as being potential agents of carcinogenesis. But maybe tumors can be initiated by simple repetitive heating of tissue; no one is sure about this, and prudent people would be wise to use wired headpieces when using cell phones.

Skiers and snowboarders are renowned for leaving the slopes with a sunburn. But the month matters mightily. Surrounding yourself with snow in March, when the Sun is much higher, will produce a burn three times faster than it will in December.

When it comes to escaping from the Sun's surface, the rules are inflexible. An object must be boosted to a speed of at least 384 miles per second, or it will not be able to overcome the Sun's enormous gravity.

CHAPTER 11: THE SUN BRINGS DEATH

If 98 percent of melanoma deaths occur among whites, why is mortality so high among the relatively few nonwhites who get this disease? Health costs and economics no doubt play a role. In addition, perhaps, blacks in particular may not think of melanoma as something that can happen to them, and hence they dismiss odd skin lesions.

Although most experts say that 65 percent of melanomas are linked to UV exposure, some say that the connection may be as high as 90 percent. There is currently no unambiguous way to know for sure.

CHAPTER 12: THE SUN WILL SAVE YOUR LIFE

Autism caused by lack of vitamin D? See Gabrielle Glaser, "What If Vitamin D Deficiency Is a Cause of Autism?," *Scientific American*, April 24,

2009, and the 2008 study by Swedish researchers in *Developmental Medicine and Child Neurology.*

John Cannell's primary vitamin D focus is on its putative link with autism. He is convinced that a serious D deficiency is a major trigger for that heartbreaking affliction, and he has helped initiate new studies to see whether mothers of autistic children, who have a 10 percent chance of having another autistic child, can drastically reduce that rate by taking vitamin D supplements. He told me, however, that he must often battle autism organizations that are so convinced that vaccinations are the cause, they regard him as an enemy, a distraction from the "real" crusade.

CHAPTER 13: I'M AN AQUARIUS; TRUST ME

Another major planet grouping happened on May 5–8, 2000, with no noticeable earthly effects either. Indeed, such conjunctions occur once a decade or so on average. Thanks to computers, it is easy to search for correlations between historical planetary groupings and earthly events of many kinds. The results, not surprisingly, are random, meaning negative. By contrast, the Sun, because of its mass, and the moon, because of its nearness, cause measurable tidal stresses on our planet. Though less than 0.1 percent as strong as the stresses already present in crustal rock, they are apparently enough to occasionally trigger earthquakes that are "on the brink." Earthquake onset thus shows a very small but nonzero correlation with spring tide configurations (when the moon and Sun act together and "pull" in the same direction), particularly when the moon is near its perigee.

The book that predicted a California calamity based on the Sun's reaction to a planetary alignment is *The Jupiter Effect* by John Gribbin and Stephen Plagemann (New York: Walker, 1974).

Yearly back-and-forth oscillations caused by our planet's tilt and elliptical orbit do vary the daily local sunrise and sunset times, especially for those at high latitudes. We're all very aware that sunrise and sunset do not happen at the same time throughout the year. But the sunrise does not migrate around the clock; the Sun always manages to come up in the morning. In most places, the sunrise deviates from its average time by only ninety minutes. Changing the clocks for daylight saving time adds another hour, of course.

Astrologers in Europe and the United States use imaginary "signs" and not the actual stars and constellations, so they place the New Year's Sun in the sign of Capricorn. In the actual night sky, however, the Sun on January 1 hovers just above the famous "teapot" of Sagittarius.

Why use day or time of *birth*? If there is to be any sort of solar connection or influence, why not the Sun's position at the time of *conception*? Are we to believe that if the doctor's golf game delayed his performing a C-section for two hours, thus changing all the "houses" of one's natal horoscope, the baby's entire personality would be dramatically and forever altered? This is a rhetorical question, because of course we know why moment of conception is not used for reckoning anything, as countless paternity suits have attested.

Statistical studies involving marriage and divorce are very easy and inexpensive to perform. Marriage and divorce records are readily available (part of the public record), a huge amount of data can be garnered quickly, and no one has to be paid to participate.

CHAPTER 14: RHYTHMS OF COLOR

Why are paint's primary colors different from those of light? Using light's primary colors blue, green, and red in various combinations, one can create every possible color; this is how a TV works. Adding more colors always makes an image brighter, since the process of combining means supplying extra light. At a certain point, we reach white.

Paint, by contrast, is *subtractive.* A pigment does not glow on its own; we see paint only because external light (presumably white light, which contains all colors, or at least red, blue, and green light) is hitting it. The pigment absorbs some colors and reflects others. Since this process always involves absorption, additional paint prevents further reflections, leaving even less light to reach our eyes: the image darkens. Add too many colors and the result is muddy or black, because now nearly all light is absorbed by the paint's molecules and nothing is reflected.

Interestingly, mixing any two primary colors of light produces one of the primary colors of paint. That is, blue and green light produce cyan paint; blue and red make magenta; and green and red produce yellow. With these pigments—yellow, cyan, and magenta—an artist can create all other colors. Here's how it works: Yellow paint looks that way because its chemical absorbs white light's blue component but reflects red and green to our eyes—and red and green light always subjectively combine to appear yellow. Cyan paint absorbs red but reflects blue and green. So when yellow paint is mixed with cyan, the combination absorbs both blue and red light; the only color that both pigments reflect is green.

Thus, an artist can create green *paint* by mixing cyan and yellow, but green *light* can never be created with any kind of mixture. TV sets and the Sun display green only when pure green light is present. Light and paint, therefore, do not merely have different primary colors; the colors created by their combinations are accomplished by entirely different processes as well.

The bouncing of sunlight among air atoms and molecules is an example of Rayleigh scattering (see chapter 10), first described two hundred years ago by Lord Rayleigh (John William Strutt). This phenomenon varies with the fourth power of wavelength, which means shorter waves such as violet scatter much more than longer wavelengths such as red. Red sunlight, at 650 nanometers, scatters only one-fourth as much as blue sunlight, at 450 nanometers. Violet, at 400 nanometers, scatters even more, but the human eye barely sees this color, which is why we do not perceive the true violet hue of the daytime sky. Still, all colors scatter to some degree, so that a handheld spectroscope pointed at the daytime sky reveals not just the dominant blue but all of the Sun's rainbow colors.

Nineteenth-century astronomers, eager to be the first to detect new galaxies or nebulae, supposedly sometimes used belladonna to enlarge their pupils beyond the normal 7 mm limit in an effort to let in a superhuman amount of light and brighten objects seen through their telescopes.

CHAPTER 15: PARTICLE MAN

Roger Waters of the rock group Pink Floyd has said that he "borrowed" these lyrics from the book *Poems of the Late T'ang* translated by A. C. Graham.

My own legendary clumsiness makes me relate fondly to Wolfgang Pauli's reputation for supernaturally unrelenting lab accidents and broken equipment, which became known as the *Pauli effect.* This, of course, should not be confused with the *Pauli exclusion principle,* which remains one of the physicist's greatest discoveries and a fundamental truth of quantum mechanics, and has nothing to do with clumsiness. It essentially says that two particles with the same properties (such as electrons having the same "spin") cannot share the same orbital location in an atom.

The value of alpha, the strength of the electromagnetic interaction, which is a way of expressing the way light is created, remains disquieting even today. Decades after Pauli's heyday, the great American physicist Richard Feynman, writing in 1985, famously said of it:

> It has been a mystery ever since it was discovered more than fifty years ago, and all good theoretical physicists put this number up on their wall and worry about it. Immediately you would like to know where this number for a coupling comes from: is it related to π or perhaps to the base of natural logarithms? Nobody knows. It's one of the greatest damn mysteries of physics: a magic number that comes to us with no understanding by man. You might say the "hand of God" wrote that number, and

> "we don't know how He pushed his pencil." We know how to experimentally measure this, but we don't know what kind of dance to do on the computer to make this number come out, without putting it in secretly!

Writer Warren Johnson said of it, "The reason for 1/137 is much deeper than some coincidence with angles, number sequences or even religions. It is very likely that the fine structure constant was this finely tuned for some final end (to put it Socratically and teleogically) not yet realized by humans. Whether the human race is the reason/final end, or just a step on the journey only time will tell."

The Homestake Mine experiment brought scientific precision to a new and almost unbelievable level. Out of the 10^{30} atoms of cleaning fluid—about the same number as expressing the weight of the Sun in pounds—Davis was able to detect a few dozen of the neutrino-created argon atoms, after waiting a few months for the newly hatched argon to accumulate. It was an amazingly precise experiment, and yet the detected quantity was smaller than the expected number by a factor of two or three, which disappointed everyone.

You thought nothing could go faster than light? That's true—in a vacuum such as space. But in any denser medium, such as air or water, light moves more slowly, and therefore some particles can outrun it under those conditions. The blue glow common in nuclear reactor cores is Cerenkov (also spelled Kerenkov and Cherenkov) radiation, a sign that electrons are "breaking the light barrier" in those pools of water.

CHAPTER 16: TOTALITY

How much time do we have left to enjoy total solar eclipses? The answer varies with the authority. One problem is that the rate at which the moon is leaving us will not always be the same as it is today. We're certainly safe for the next few million years and maybe as many as 100 million.

For those who are curious as to how long one must wait between total solar eclipses, here is a list of selected North American cities and the current interval between totalities.

Location	Most Recent Totality	Next Totality	Years Between Eclipses
Anchorage, AK	1943, February 4	2399, August 2	457
Atlanta, GA	1778, June 24	2078, May 11	300
Boston, MA	1959, October 2	2079, May 1	120
Chicago, IL	1806, June 6	2205, July 17	399
Dallas, TX	1623, October 23	2024, April 8	402
Denver, CO	1878, July 29	2045, August 12	167
Halifax, NS	1970, March 7	2079, May 1	109
Honolulu, HI	1850, August 7	2252, December 31	402
Houston, TX	1259, October 17	2200, April 14	941
Las Vegas, NV	1724, May 22	2207, November 20	484
Los Angeles, CA	1724, May 22	3290, April 1	1,566
Mazatlán, Mexico	1991, July 11	2024, April 8	33
Mexico City, Mexico	1991, July 11	2261, December 22	270
Miami, FL	1752, May 13	2352, February 16	600
Montreal, QC	1932, August 31	2024, April 8	92
New Orleans, LA	1900, May 28	2078, May 11	178
New York, NY	1925, January 24	2079, May 1	154
Phoenix, AZ	1806, June 16	2205, July 17	399
St. Louis, MO	1442, July 7	2017, August 21	575
San Francisco, CA	1424, June 26	2252, December 31	829
Seattle, WA	1860, July 18	2645, May 17	785
Toronto, ON	1142, August 22	2144, October 26	1,002
Washington, DC	1451, June 28	2200, April 14	749
Woodstock, NY	1925, January 24	2079, May 1	154

CHAPTER 17: THAT'S ENTERTAINMENT

If you head to Alaska in March to see the northern lights, what conditions can you reasonably expect? During the three March fortnights I lectured on the aurora there, we usually had 15°F days and –20°F nights.

One year, however, daytime highs were more like 10°F, but the staff at Chena Hot Springs, ninety minutes east of Fairbanks, supplied arctic clothing at no charge.

Full disclosure: our "northern lights alert" phone chain was frequently a pain in the neck. I'd reluctantly dash indoors to get to the phone (no cell phone service in these mountains, even now), only to miss the display while dialing people at 1:00 AM, and often getting an annoyed "Who is this?" when the other party picked up. When I explained the purpose of the call, the grumpiness persisted, since many of the people who'd signed on had neglected to inform their significant ogres of the possibility of a postmidnight wake-up call. I would be interested in hearing how your "aurora hotline" turns out, for a future article. Please send reports to my website, http://skymanbob.com/.

You can find a great image of a circumzenithal arc (CZA) at http://www.atoptics.co.uk/halo/cza.htm.

Supernumerary arcs are created by a different process than rainbows—namely, diffraction rather than refraction. Here, instead of each color being bent to fly away in a slightly different direction, the light waves interfere with one another. It is proof that light is a wave and not a particle, or at least is acting that way at the moment.

For a good photo of Alexander's dark band, go to http://www.atoptics.co.uk/rainbows/adband.htm.

Although it's true that you can't see a reflected rainbow and simultaneously the rainbow itself, check out this website—http://www.atoptics.co.uk/rainbows/rflctd.htm—for a look at an *apparent* reflected rainbow in a lake and a good illustration of what is actually going on.

For some good images of Sun dogs, along with the 22-degree halo, go to http://www.atoptics.co.uk/halo/parhelia.htm.

Have a cheat and look at cloud iridescence at http://www.atoptics.co.uk/droplets/irid1.htm.

CHAPTER 18: COLD WINDS

The height of Everest remains controversial, as with everything else in the world that you'd think had been settled. The official Nepal and China listings (the countries on whose borders the mountain sits) say 29,029 feet, which is an oft-quoted elevation. A more recent measurement from October 2005 gives an "official" height of the loftiest rock, excluding snow and ice, as 8,844.43 meters (29,017.16 feet), with a claimed accuracy of ±0.21 meter (0.69 foot). The Swiss Foundation for Alpine Research disagrees. In May 1999, a US Everest expedition directed by Bradford Washburn attached a permanent GPS unit to the mountain's highest bedrock, where it continues to give an elevation of 8,850 meters (29,035 feet), with a snow/ice elevation 1 meter (3 feet) higher. This figure, too, is often cited. Whatever the height, Everest has been cited as the terrestrial point closest to the Sun.

But it is not! Thanks to Earth's tilt and Everest's latitude of 28 degrees north, the Sun is never directly above Everest. This, plus Earth's equatorial bulge, means that even low-lying regions in the tropics are closer to the Sun than Everest is.

Confusingly enough, "glacial periods" are colloquially and commonly called "ice ages," but here we use the latter term as scholars and experts do, to refer exclusively to Earth's vastly longer cold epochs.

If humans didn't meddle with the climate, what would our planet be like a few thousand years hence? Warmer? Colder? Until recently, experts believed that we're now finishing the current interglacial and will enter the next glacial period in only a couple of thousand years. But if you trust Milankovitch, as I and an increasing number of researchers do, then the next glaciation won't arrive for 50,000 and possibly 130,000 more years.

Where would you get your best view of the starry universe — from down here on the ground or from space? Space shuttle commander Andy Thomas, who grew up in the Australian outback and thus knows the pristine starry heavens better than most of us, told me an amazing thing. Fewer stars are visible out the International Space Station (ISS) windows than from an unpolluted rural location on the ground. That's because our atmosphere dims stars by a barely detectable one-third of a magnitude, less than the absorption of the ISS windows.

Three billion years ago, the Sun emitted 30 percent less energy than it does today. At that time, each square yard of air atop equatorial Earth received only about 900 watts of solar irradiance. So why wasn't Earth frozen solid? This conundrum puzzled experts for a long time, but they were probably overthinking the issue. Obviously, back then Earth had a different atmosphere with a lot more carbon dioxide and plenty of methane, which trapped so much heat that temperatures were pretty much the same then as now.

Most solar experts think that cycle 24 will peak around May 2013 and that it will exhibit another wimpy maximum that might be even weaker than cycle 23's.

Callendar's numbers were mostly spot-on, and it was he who first documented a rise in global temperatures as well as explained it. The first person to have predicted that carbon dioxide from the burning of fossil fuels would cause global warming, however, was probably the Swedish scientist Svante Arrhenius, in an 1896 paper titled "On the Influence of Carbonic Acid in the Air upon the Temperature of the Ground." But Arrhenius thought it would take humans three thousand years to double the air's CO_2. The actual time period, unfortunately, is closer to one hundred years.

The critical issue of the ocean's ability to absorb carbon dioxide is complex. There is certainly enough ocean water to keep taking in half our human carbon output forever. The problem is that only the uppermost sea layers are exposed to the air, and it is these layers that are becoming increasingly saturated and acidic, due to carbonic acid. The oceans are thus unable to absorb as much as they did fifty or even ten years ago. The lower, pristine ocean levels do eventually rise and mix up to the surface, but it's a slow process. Oceanographers debate how long it takes for the top one hundred or three hundred feet of the sea to get entirely replaced by the deeper waters below. Half a dozen decades? More than a century? Adding to the fun, the time varies around the world according to ocean temperature stratification and undersea currents.

With all four factors added together, the matchup with temperature variations is excellent but not perfect. Those thirty years of global temperature flatness starting at the end of World War II are still a bit weird.

There is still some controversy regarding the Sun's influence on the climate. Some people say the Sun lost its dominance between 1975 and 1980; others say 1994.

CHAPTER 19: THE WEATHER OUTSIDE IS FRIGHTFUL

Richard Carrington was the first human being ever to report observing a solar flare, some 150 years ago.

A coronal mass ejection (CME) is a dangerous event. A typical CME lets loose ten billion tons of subatomic particles, traveling at 1/200 the speed of light, like a swarm of bees on meth. This easily penetrates skin, bone, spacecraft, anything. A strong CME could kill astronauts on the moon or on Mars, or en route. Shannon Lucid, America's most experienced woman astronaut, is far from alone in saying that this is the greatest danger facing future manned spaceflight. CMEs are not just waves of energetic light, such as gamma rays. Rather, CMEs are bullets. They tear cells apart and damage DNA. Astronaut Alan Shepard openly mused that the leukemia that eventually killed him might have come from the radiation he received during his *Apollo 14* mission to the moon. Had his flight been launched just before a CME blasted from the Sun—as very nearly occurred during the later *Apollo 17*—he wouldn't have had the chance to wonder. The crew would unquestionably have been killed.

My brief summary of the fundamentals of space weather events comes from the booklet *Severe Space Weather Events—Understanding Societal and Economic Impacts: A Workshop Report,* prepared by the Committee on the Societal and Economic Impacts of Severe Space Weather Events: A Workshop (Washington, DC: National Academies Press, 2008).

To get the current readouts of the solar wind's magnetic polarity, speed, density, and more, go to http://www.swpc.noaa.gov/ace/MAG_SWEPAM_24h.html. The info looks wonderfully high-tech, and you will impress anyone looking over your shoulder. If your boss suspects that you might be downloading from salacious websites, always have this at the ready; it will cure him of any such notion. Here's how to read the graphs: The topmost graph — magnetic polarity — is very important, because incoming solar particles will interact with us only if their magnetic field has its south pointing up, opposite ours. This is indicated by the trace marked "BZ." You can count on the incoming particles affecting us when the magnetic trace line is in the upper part of that graph, above the horizontal midpoint line. If you're trying to predict an aurora, also look at the middle graph, the speed of the solar particles. You want to see a speed above 700 kilometers per second and a high density, with more than 20 or 30 particles per cubic centimeter. During a true CME, all these readings will go berserk. That's when you should get in the car and go visit those friends in the country. Remember, ACE gives about two hours' warning. Don't hesitate.

CHAPTER 20: TOMORROW'S SUN

The Ring Nebula gives us a colorful peek at what the Sun and its immediate environs will look like six billion years from now. Go to http://apod.nasa.gov/apod/ap060625.html.

Water's weight depends on its temperature. A full aquarium one foot on each side weighs sixty-two pounds if the water is 100°F (in which case the fish had better be very tropical) but half a pound more if the water is just under 40°F, which is the temperature of water's maximum density.

The white dwarf and then black dwarf Sun will actually boast even a greater percentage of carbon and oxygen than the 20 percent in a human exhalation. An out-breath contains about 16 percent oxygen and 4 percent carbon dioxide.

Bibliography

BOOKS

Brody, Judit. *The Enigma of Sunspots.* Edinburgh: Floris Books, 2002.

Dick, Thomas. *Celestial Scenery, Sidereal Heavens.* Philadelphia: E. C. & J. Biddle, 1847.

Espenak, Fred. *Fifty Year Canon of Solar Eclipses, 1986–2035.* NASA Reference Publication 1178. Washington, DC: NASA, 1987.

Golub, Leon, and Jay Pasachoff. *Nearest Star.* Cambridge, MA: Harvard University Press, 2001.

Greenler, Robert. *Rainbows, Halos, and Glories.* Cambridge: Cambridge University Press, 1980.

Harwit, Martin. *Astrophysical Concepts.* 2nd ed. New York: Springer-Verlag, 1988.

Maran, Stephen P. *The Astronomy and Astrophysics Encyclopedia.* New York: Van Nostrand Reinhold, 1991.

Maunder, Annie, and Walter Maunder. *The Heavens and Their Story.* London: Robert Culley, 1901.

Meeus, Jean. *Astronomical Tables of the Sun, Moon, and Planets.* 2nd ed. Richmond, VA: Willman-Bell, 1995.

Olmsted, Denison. *Olmsted's School Astronomy.* New York: Robert B. Collins, 1856.

Poppe, Barbara. *Sentinels of the Sun.* Boulder, CO: Johnson Books, 2006.

Rawicz, Slavomir. *The Long Walk.* New York: Harper & Row, 1956.

Soon, Willie Wei-Hock, and Steven H. Yaskell. *The Maunder Minimum and the Variable Sun-Earth Connection.* Singapore: World Scientific Publishing, 2003.

Weisberg, Joseph. *Meteorology.* 2nd ed. Boston: Houghton Mifflin, 1981.

CENTRAL ARTICLES

Agee, Ernest M., Emily Cornett, and Kandace Gleason. "An Extended Solar Cycle 23 with Deep Minimum Transition to Cycle 24: Assessments and Climatic Ramifications." *Journal of Climate* 23 (2010): 6110–6114.

Broecker, Wallace S., David L. Thurber, John Goddard, Teh-lung Ku, R. K. Matthews, and Kenneth J. Mesolella. "Milankovitch Hypothesis Supported by Precise Dating of Coral Reefs and Deep-Sea Sediments." *Science,* January 19, 1968.

Callendar, Guy S. "Temperature Fluctuations and Trends over the Earth." *Quarterly Journal of the Royal Meteorological Society* 87, no. 371 (1961).

D'Aleo, Joseph. "Is Global Warming on the Wane?" In *The Old Farmer's Almanac.* Dublin, NH: The Old Farmer's Almanac, 2009.

Eddy, J. A. "The Maunder Minimum." *Science,* June 18, 1976.

Jevrejeva, S., A. Grinsted, and J. C. Moore. "Anthropogenic Forcing Dominates Sea Level Rise Since 1850." *Geophysical Research Letters* 36 (October 28, 2009).

Shindell, Drew T., Gavin A. Schmidt, Michael E. Mann, David Rind, and Anne Waple. "Solar Forcing of Regional Climate Change During the Maunder Minimum." *Science,* December 7, 2001.

Trouet, Valérie, Jan Esper, Nicholas E. Graham, Andy Baker, James D. Scourse, and David C. Frank. "Persistent Positive North Atlantic Oscillation Mode Dominated the Medieval Climate Anomaly." *Science,* April 3, 2009.

IMPORTANT WEBSITES

ACE Dynamic Plots, NOAA/Space Weather Prediction Center: http://sec.noaa.gov/ace/MAG_SWEPAM_24h.html.

Atmospheric Optics: http://www.atoptics.co.uk/. A comprehensive collection of sunlight-produced diffraction and refraction phenomena—a gorgeous website.

Auroral Activity Extrapolated from NOAA POES (Polar-orbiting Operational Environmental Satellite): http://www.swpc.noaa.gov/pmap/index.html.

The Sunspot Cycle, Marshall Space Flight Center: http://solarscience.msfc.nasa.gov/SunspotCycle.shtml.

Index

About the Author

BOB BERMAN, one of America's top astronomy writers, wrote the popular "Night Watchman" column for *Discover* for seventeen years. He is currently a columnist for *Astronomy,* a host on NPR's Northeast Public Radio, and the science editor of *The Old Farmer's Almanac.* He lives in Willow, New York. Visit him at http://skymanbob.com.

BACK BAY · READERS' PICK

Reading Group Guide

The Sun's Heartbeat

And Other Stories from the Life of the Star That Powers Our Planet

by

BOB BERMAN

A conversation with Bob Berman

What got you started in astronomy?

It was destiny. My very first memory was of looking out of a carriage or stroller and seeing the starry night sky. I memorized all the named stars and their distances and spectral classes when I was a teenager, when my friends were memorizing baseball card stats. Of course, I couldn't wait to start studying it in college. But I had no idea I'd actually make my living as an astronomer.

Have you ever discovered a new comet, or anything else previously unseen by astronomers?

Hardly. Early on, I got a job writing popular science for a weekly newspaper, which led to being offered the honor of becoming *Discover*'s monthly astronomy columnist, which I did for seventeen years. I was soon offered the position of astronomy editor of the *Old Farmer's Almanac,* as well as being columnist and contributing editor of *Astronomy* magazine, and I've remained at both posts for nearly twenty years. So, although I have taught physics and astronomy at the college level, and had the fun of popularizing it for decades on a weekly radio show, my actual job description is the pleasant task of keeping up-to-date on all the latest discoveries, speaking directly to the researchers, and then translating their hard work for the general public. I do not do any research myself.

You've traveled in thirty-five countries, over a span of many years. Why did you choose to live in upstate New York?

I think the rural mountains of the Northeast are among the most beautiful places on our planet, with friendly people to boot. I live near a village where people still do not lock their doors, and where one routinely sees true wildlife, like bears and such, from one's kitchen window. I was raised in a big city, but I'm not enamored of urban life. I love the starry skies of the country, and I've lived here thirty-five years. Is it slow-paced and boring? Yes, and that's what I like.

What do you think is the greatest myth regarding popular astronomy?

The widespread notion that you need a telescope. Though I actually own a private observatory and thus have no axe to grind either way, I still regard binoculars as the best way to go. With the correct filters (and one has to be very careful here), the Sun is amazing through binoculars—and so is the Milky Way. But the naked-eye view is the best of all, if one is in an unpolluted place.

Are you looking forward to any particular upcoming eclipses?

Definitely. But there are actually two. After an unprecedented thirty-eight-year drought, the U.S. will finally get a pair of total solar eclipses, on August 21, 2017, and then April 8, 2024. To have the magic of a totality in my own country will be very special.

I hope everyone makes the effort to position themselves within those hundred-mile-wide ribbons of the moon's shadow. The precise tracks will be widely published a year or two ahead of time.

Do you have a favorite section of the sky? A particular time of year?

Each season offers its own wonders. Favorite planets telescopically are of course Jupiter and Saturn. To the naked eye, Venus is always riveting and special. But the Sun and moon, as ordinary as they may superficially seem, are both fantastic when viewed with the right wide-field equipment. As for the stars and constellations, the summer and early fall offer the Milky Way at its best, which was regarded as nothing less than the center of the universe to the ancient Meso-American cultures. That's when one also sees the glories of Scorpius and the Teapot to its left. The winter features Orion and his friends, which present the brightest collection of stars we ever see. Spring alone is rather dull in the heavens, and the Milky Way is then coincident with the horizon. But that's when we've been given the clearest window into intergalactic space, up past the Big Dipper. Good telescopes then come into play to reveal the most galaxies off in the distance, along with the great globular cluster in Hercules. There's always something good in the sky.

Is physical science getting close to "figuring it all out"?

We periodically hear that. For example, string theory was supposedly going to unify all forces and make everything clear, and the Higgs boson was supposedly going to provide the final critical jigsaw puzzle piece for particle physics. I think it's just the opposite. I feel that what I know is like a single snowflake in a blizzard compared to the still unknown. Paradoxically, the more fundamental the issue, the deeper the mystery remains. We know how the Sun and stars were born, but not whether the cosmos is finite or infinite. The biggest underlying mystery, in my opinion, is that we lack the slightest understanding of our own consciousness or how human awareness has even arisen. And since what we perceive critically depends on our nature and limitations as perceivers, this absence of information renders all fundamental cosmological knowledge virtually meaningless at this point. Thus, science is still in its infancy.

Bob Berman's solar eclipse journal

I have been fortunate enough to have visited thirty-five countries, to have traveled extensively in Asia by motorcycle, to have piloted planes and flown over snowy winter mountains in Alaska, and to have snuck up on penguins in Tierra del Fuego. But I would trade any of it for a total solar eclipse—not a lunar eclipse, and not a partial solar—because that remains the most powerful experience I have ever known.

In this book, I've tried to convey the ineffable quality of that event, when the Sun, moon, and your own body form a perfectly straight line in space. Yes, people weep and animals go crazy. But each eclipse is unique. Since it is usually located far from your home, and travel is expensive, a totality is a pilgrimage situation, and it is very important to try to choose a place that is most likely to have clear skies. Whenever I've been the "eclipse astronomer" and could give input into where the group should go, I've followed the statistical odds. But sometimes that doesn't work out. That's why the literature is filled with heartbreaking accounts of failed, clouded-over eclipse expeditions.

I have traveled to or been hired for seven total eclipses plus two annular eclipses (where a ring, or annulus, of sunlight remains around the moon) and been extremely lucky to have seen all except for one of the annulars. I offer my notebook here in the hope that it may light the fire of determination in some readers, to inspire them to suffer considerable expense and trouble in order to witness the fantastical apparition of the moon swallowing up the Sun.

March 7, 1970, Virginia Beach, Virginia:
A girlfriend and I drive all through the night, to find cloudless skies and a vast beach with many people gathered as far as the eye can see. I use a doubled-up strip of exposed and developed color film, instead of the correct black-and-white, to watch the dangerous partial phases of the eclipse, and am fortunate not to have damaged my eyesight. The three-minute totality is astonishing. Beyond words. It far exceeds my already lofty expectations. Though I was still an adolescent, it was life-altering.

February 16, 1980, Konârak Sun Temple, Orissa State, India:
Very few people gather on the sands surrounding this amazing, colossal thirteenth-century monument to the Sun. It is almost deserted—probably due to the superstitious, fearful populace. I am here with an Indian astrophysicist and an Indian newspaper reporter. I meet a couple from the Midwest, who sit a short distance from us, observing the event. After the four-minute totality they are so moved, they can barely speak. This was the most magical, powerful, and touching totality for me.

May 30, 1984, southeastern United States:
We turn our car around. It is raining over the entire eclipse path. It's hopeless. Oh, well. At least it's an annular eclipse, and not total.

July 11, 1991, Cabo San Lucas, Baja, Mexico:
At six and a half minutes, this is the longest totality for the entire next century. I have been hired as an "eclipse astronomer" for a

large tour group. This is statistically the clearest place in North America in the month of July, which is why I chose this one instead of Hawaii, where it will also be total (it was ultimately clouded-out there). But an overcast covered the sky the day before this event at eclipse time, which was worrisome. On the eleventh, a high thick overcast starts slowly approaching from the northeast, but does not reach the Sun during the event. The eclipse goes on and on, with the sky unusually dark, and many stars appear. Unfortunately, two members of our group, who rented a car to go forty miles up the beach to gain a few extra seconds of totality, were clouded-out from their location. Also sadly, another member of our group broke his neck in the extremely rough surf and had to be airlifted out.

May 10, 1994, upstate New York:
The path of annularity sweeps right over Albany, practically in my backyard. But I accept a lecture job instead at the Victorian Mohonk Mountain House, a surreal National Historic Landmark, where the Sun will be 99 percent covered. I expect very little on this perfect, cloudless day, since I regard partial eclipses as ho-hum affairs that lack the magic (and the corona, prominences, and darkness) of a totality. Instead, I am surprised and delighted that the eerie lighting over the scene, with high contrast and polarized light, is identical to the experience immediately before and after a full totality. It's a winner, after all. Clearly, a 99 percent partial eclipse still offers treasures. The crowd is awed.

February 26, 1998, 50 miles off the coast of Colombia:
I am again the "eclipse astronomer," this time traveling through the Panama Canal aboard a Cunard science cruise—the sole westernmost ship for this totality. Most travelers went to islands in the Caribbean to see this, but we're going to where the totality will

last the longest, a full four minutes. This is my first shipboard eclipse, and I see the advantage: Despite generally cloudy Pacific skies, the navigator, captain, and I, using satellite guidance, find a two-hundred-mile-wide hole and stop the engines under clear skies. This event and the last one were fabulous, but somehow lack the tear-inducing magic of my first two totalities. Is the wildly varying shape of the corona, due to the different levels of solar storm activity during the eleven-year solar cycle, playing a role in the drama?

August 11, 1999, on the Black Sea off Turkey:
Again I am the "eclipse astronomer"—lecturing for the group put together by the great George Schy. Many try to see this totality from southern England, northern France, and Germany, where it winds up being completely overcast and rainy. Here (and in Turkey farther east) the skies are clear for the two-minute, twenty-second totality. I request that the crew set me up in the crow's nest, high above everything, and they run what seems like a quarter-mile microphone cable hooked up to the ship's loudspeakers. I stop my discourse twenty minutes before totality, since I firmly believe such a sacred experience merits silence and not human narration. Meanwhile, another ship has arrived to drop anchor just a half mile from ours, apparently thinking, "This must be the place." We can faintly hear that someone on *that* ship is blabbering through the whole thing. He should be shot. After totality ends, all ships blow their horns, which is deafening to me and my family in the crow's nest. The crew seems intoxicated.

March 29, 2006, northwestern Egypt:
The Libyan border is in sight in the distance as our forty-five-member tour group gathers on the desert sands for the four-minute totality. I completed my final lecture the night before. This

morning I've guided our bus down the highway, away from the crowds and raucous music of the government-created "eclipse site," in which President Mubarak has arrived by helicopter. It is foggy this morning, but it thankfully burns off an hour before the eclipse begins. I get lucky once again. It is, as always, spectacular. Several people weep.

November 13, 2012, Cairns, Australia:
I am once again to be the "eclipse astronomer" for George Schy's travel group. This time the Sun will only be thirteen degrees high — a perilous situation so far as being clear of clouds is concerned, especially since the dry season has ended. When skies are partly cloudy, large breaks of blue commonly appear overhead, but clouds tend to "fill in" when they're viewed sideways, closer to the horizon, and that's where the Sun will be situated for this early morning event. Statistically, we have a 55 percent chance of seeing this two-minute totality. It's a coin toss. I had a friend who has been clouded-out of four of the seven eclipses he has traveled to see. Will my good fortune hold out yet again?

Questions and topics for discussion

1. If you were a member of a primitive culture, with no instruments but merely the power of observation, what would you hypothesize about the Sun?

2. What is the most astonishing new fact you learned about the Sun from reading *The Sun's Heartbeat*?

3. Why was the discovery of sunspots so important? What did it teach us?

4. A mere two centuries ago, brilliant minds concluded that there was no possible way to ever learn the Sun's true composition. Which invention instantly changed that?

5. After he discovered the planet Uranus, what was William Herschel's most humorously outrageous claim about the Sun?

6. One eighteenth-century writer claimed the Sun was made of ice. What was his reasoning? What would you think the Sun was made of if there wasn't anyone to tell you otherwise?

7. Earth's magnetic field "flips" and reverses itself three times each million years. The Sun's north and south poles trade places too—but with what frequency? Did you imagine this was possible before you read *The Sun's Heartbeat*?

8. If you could witness a single dramatic Sun-related spectacle, what would it be?

9. How did disobeying his superiors (and getting fired) wind up earning Pieter Zeeman the Nobel Prize in Physics? Would you have done the same?

10. What is actually "blowing" in the solar wind that pushes on comet tails to make them point away from the Sun?

11. In your opinion, what is the most medically dangerous entity that arrives here from the Sun?

12. Do you think you should let your skin receive more direct sunlight than you've been getting, or less? Did reading this book change your perspective on that? Explain why.

13. A total solar eclipse is widely said to be an awe-inspiring experience. Do you ever plan to travel to see one? Do you have a particular future eclipse in mind?

14. Some researchers think the Sun may be entering an extended period of emitting less light. Do you think this reduced brightness could counterbalance global warming?

15. What fact about astronomy were you the most surprised to learn while reading *The Sun's Heartbeat*?

ALSO AVAILABLE FROM BACK BAY BOOKS

The Disappearing Spoon

And Other True Tales of Madness, Love, and the History of the World from the Periodic Table of the Elements

by Sam Kean

"Ebullient. . . . A nonstop parade of lively science stories."

—Janet Maslin, *New York Times*

"Irresistible. . . . Kean gives science a whiz-bang verve so that every page becomes one you cannot wait to turn just to see what he's going to reveal next." —Caroline Leavitt, *Boston Globe*

"Kean unpacks the periodic table's bag of tricks with such aplomb and fascination that material normally as heavy as lead transmutes into gold." —Keith Staskiewicz, *Entertainment Weekly*

"Great good fun. . . . A lively and edifying history of the elements. . . . An engrossing, adventurous read."

—Leonard Cassuto, *Salon*

Back Bay Books
Available wherever paperbacks are sold

ALSO AVAILABLE FROM BACK BAY BOOKS

Just Food

Where Locavores Get It Wrong and How We Can Truly Eat Responsibly

by James E. McWilliams

"Enlightening. . . . *Just Food* offers a brave, solid argument that anyone who cares about their food should consider."

—Meridith Ford Goldman, *Atlanta Journal-Constitution*

"Everyone who has read up on their Michael Pollan should also read *Just Food.*" —Russ Parsons, *Los Angeles Times*

"Bold. . . . *Just Food* makes a challenging case for changing the way we feed the world." —Kim Pierce, *Dallas Morning News*

"McWilliams challenges his readers' preconceptions with a fair-minded review of the evidence concerning the environmental impact of our food choices."

—Jacob Sullum, *Wall Street Journal*

Back Bay Books
Available wherever paperbacks are sold

ALSO AVAILABLE FROM BACK BAY BOOKS

Crow Planet

Essential Wisdom from the Urban Wilderness

by Lyanda Lynn Haupt

"With her sensitivity, careful eye, and gift for language, Haupt tells her tale beautifully, using crow study to get at a range of ever-deepening concerns about nature and our place within it, immersing us in a heady hybrid of science, history, how-to, and memoir."

—Erika Schickel, *Los Angeles Times*

"In a lyrical narrative that blends science and conscience, Haupt mourns the encroachments of urbanization, but cherishes the wildness that survives."

—Liesl Schillinger, *New York Times Book Review*

"An inspired meditation on our own place in nature. . . . You will never look at crows in the same way again."

—Jenifer Rhoades, *Washington Post*